Advance Praise for *Comprehending the Climate Crisis*

"Dr. Dibble's book is an excellent overview of climate change. Covering basic chemistry and cosmology—and taking an historical approach that stretches back before the Industrial Revolution—he explains the science that's so necessary to truly grasping this issue. Readers will come to understand global warming's devastating effects but they won't leave with hopelessness. Dr. Dibble sets out practical solutions—like ending coal-fired power and ramping up solar and wind energy—that in relatively short order could bring great improvement."

—Gideon Forman, Executive Director,
Canadian Association of Physicians for the Environment

May 8/13

COMPREHENDING THE CLIMATE CRISIS

Everything You Need to Know about Global Warming and How to Stop It

Bradley J. Dibble, MD

To Lindsay,
Thank you for doing your part to help care for our planet.

Bradley

iUniverse, Inc.
Bloomington

Comprehending the Climate Crisis
Everything You Need to Know about Global Warming and How to Stop It

Copyright © 2011 Bradley J. Dibble, MD

All rights reserved. No part of this book may be used or reproduced by any means, graphic, electronic, or mechanical, including photocopying, recording, taping or by any information storage retrieval system without the written permission of the publisher except in the case of brief quotations embodied in critical articles and reviews.

iUniverse books may be ordered through booksellers or by contacting:
iUniverse
1663 Liberty Drive
Bloomington, IN 47403
www.iuniverse.com
1-800-Authors (1-800-288-4677)

Because of the dynamic nature of the Internet, any Web addresses or links contained in this book may have changed since publication and may no longer be valid. The views expressed in this work are solely those of the author and do not necessarily reflect the views of the publisher, and the publisher hereby disclaims any responsibility for them.

Any people depicted in stock imagery provided by Thinkstock and iStockphoto are models, and such images are being used for illustrative purposes only. Certain stock imagery © Thinkstock, © iStockphoto.

Grateful acknowledgement is made to the following to reprint previously published material in Pale Blue Dot by Carl Sagan: Copyright © 1994 Carl Sagan. Reprinted with permission from Democritus Properties, LLC. All rights reserved. This material cannot be further circulated without written permission of Democritus Properties, LLC.

ISBN: 978-1-4620-4519-8 (sc)
ISBN: 978-1-4620-4520-4 (hc)
ISBN: 978-1-4620-4521-1 (e)

Library of Congress Control Number: 2011914232

Printed in the United States of America

iUniverse rev. date: 11/3/2011

For Katherine,
who taught me how to care for my planet,

and for Matthew and Jamie,
the two most important reasons I want to.

> You don't inherit the planet from your parents;
> you borrow it from your children.
> —ancient Native American proverb

Contents

Preface ix

Section 1   The Background   1

1. Earth, Its Atmosphere, and Where Carbon Came From . . . 3
2. The Carbon Cycle: How Mother Nature Circulates Carbon and Generates Fossil Fuels 21

Section 2   The Problem   41

3. Harnessing Energy and How Fossil Fuels Are Used 43
4. Greenhouse Gases and How They Affect Our Planet 59
5. Global Warming and Its Devastating Effects 79

Section 3   The Solutions   101

6. Reducing Greenhouse Gas Emissions: The Small Steps You Can Take 103
7. Reducing Greenhouse Gas Emissions: The Big Steps Society Can Take 124
8. Progress Is Slow: Understanding the Resistance to Change . 149

Epilogue 167

Acknowledgments 175

Glossary 179

List of Illustrations 187

Index 189

About the Author 195

Preface

You may be wondering why a physician would write a book on global warming and the climate crisis. That's an excellent question that deserves an answer. Right off the bat, I'll point out that my profession has no direct bearing on how this book came to be. More than anything, I am simply a concerned citizen who also happens to be a physician. That said, however, I do believe those who have dedicated themselves to a career in medicine might have something unique to offer to the efforts at tackling environmental issues, but I'll explain that more a little later.

First, I'll start with the plain and simple truth. This book is the result of a personal journey. Like many of you, I care about the environment. I want to leave behind a planet that my children and grandchildren can thrive in. I drive a hybrid, do my best to turn off the lights when leaving a room, purchase energy-efficient appliances, and shop for local food. And like many people, I was particularly inspired after seeing the film *An Inconvenient Truth* with former vice president Al Gore. It managed to bring an unprecedented level of information to the mainstream consciousness, and ultimately, Gore won both an Oscar for the film and a Nobel Peace Prize for his efforts in tackling the problem. To me, the issues seemed self-evident. Anyone who didn't agree that this was a problem that needed a solution either was unaware of the facts and thus denying the science involved or perhaps had a personal interest in those industries dependent on fossil fuels, so they were putting personal profits ahead of doing what's right to help the planet.

But some of the criticisms against going green didn't always seem completely outlandish. Sure, nuclear power is a source of electricity that doesn't add greenhouse gases to the atmosphere, but what about the problem of nuclear waste disposal? And didn't I read somewhere that the carbon footprint of manufacturing a wind turbine is pretty sizable and that it takes many years of use before it finally becomes carbon-neutral, let alone green? On top of that, those wind turbines were reported as being effective only when positioned in reliably consistent and strong sources of wind; this means they have to be remotely located, away from most urban areas, which isn't very practical on a large scale. And if pursuing alternate sources of energy was the right thing to do, why weren't governments taking immediate action to effect the necessary changes? Why hadn't every country been embracing the Kyoto Protocol without question? Why was it an ultimately unsuccessful struggle to reach a useful accord in Copenhagen in 2010? And why did all of the suggested solutions to help—ideas such as clean coal, carbon capture and storage, and renewable sources of energy—seem to have counterarguments against them? Some of these counterarguments were admittedly a bit far-fetched, but certainly not all of them.

To better educate myself on the complexities of these issues, I started to explore these subjects in greater detail. There are a number of books on the topic of global warming and the climate crisis, with new ones being published every week, it seems. But every one I read touched only on certain aspects of the bigger picture. It was easy to find books on the science behind greenhouse gas emissions and climbing carbon dioxide levels, but these were usually written in exhaustive detail beyond the interest or understanding of the average person. Other publications simply listed what we need to do to reduce our carbon footprint without explaining the science behind why we should even consider it. Fewer still seemed to address the aspects of how making these changes would affect the global economy, either positively or negatively. Interestingly, I was never able to find a book that gave a comprehensive overview of all aspects of the issues at hand, especially in a manner that was easily within the grasp of the average reader and not only those who already had a working knowledge of the science involved. Where was the book a concerned citizen could go to in order to learn the basics?

Since I couldn't find such a book, leading me to the conclusion that perhaps it didn't exist, I decided that maybe I could do something to help fill that void. And so I did, and you are now holding in your hands the result of that decision. This book was written with the assumption that you want to know the pertinent issues. My goal is to arm you with a working knowledge on the broad aspects of this complicated subject, allowing you to participate in meaningful discussions around these topics without having to be a scientist or employed in an industry connected to energy. I wanted to write a book that could be read over a weekend, ideally from beginning to end, but could also be scanned in sections if only certain topics were of particular interest or for reference.

As a strong believer in science, I trust what the facts tell me. Those who deny that carbon dioxide levels are increasing or dispute that such a rise is contributing to increasing global temperatures will probably not get as much out of this book as those with an open mind. In my experience, skeptics are a difficult group to persuade away from their beliefs. As US Senator Daniel Patrick Moynihan once said, "Each of us is entitled to his own opinion, but not to his own facts." To me, the facts speak for themselves, and for the sake of completeness, they are reviewed here so that you can understand them clearly as well.

While trying to learn about the issues surrounding the climate and global warming, I ended up exploring many different areas in science: chemistry, biology, geology, cosmology, and physics. To better understand the hurdles to solving some of these problems, I also had to learn about political science, the workings of government, economics, and psychology. To ensure that you too have useful information on all of the concepts pertinent to global warming and the climate crisis, this book leads you through some aspects of these branches of knowledge as well, both the scientific and the socioeconomic. However, I promise that nothing is too complicated, and everything is presented in a simple and straightforward manner so that you can appreciate every aspect of how these areas pertain to the climate crisis, even if you've never studied any of them previously. At the end of each chapter I list what I consider to be the key concepts that were covered. I also list some suggested books for further reading if you wish to explore any of those concepts in greater detail. Any references within each chapter are listed there as well.

This book is divided into three sections. Section 1 primarily deals with the background science, covering where carbon originated, how it found its way on Earth, and how it got into the fossil fuels that have become the backbone of the global economy. It also describes the composition of the atmosphere before that point in time when our civilization's influence began to alter it—that is, the era prior to the Industrial Revolution of the eighteenth century. This section also takes you through some basic scientific concepts by way of examples that will help you to more easily understand topics discussed later in the book.

Section 2 covers the mechanics of the combustion of fossil fuels and how our dependence on them has led to significant changes in the atmosphere's composition along with the results of those changes, both observed and predicted.

Section 3 addresses some of the solutions that are available to consider, for individuals and families as well as for societies and governments. It stands to follow that there are hurdles to embracing these solutions, and these will also be addressed. By the time you finish section 3, you will have a good understanding of what is meant by "sustainable development," a term that I believe will help define the twenty-first century but that has yet to become part of the mainstream vocabulary.

As I stated earlier, the origins of this book have little to do with my profession. As I went through the process of writing it, however, I came to realize that as a physician, and as a cardiologist in particular, I might have some qualifications that I can offer to help educate the public about the issues at hand. First of all, I have a broad interest and education in science, and this has helped me to embrace much of the background story of the climate crisis. I find that carbon and its important role on our planet is a frequently neglected topic in general books on global warming. Second, as someone who frequently has to explain complicated subjects to patients in a way that helps them understand well enough to make informed decisions about the next steps they should take, I've become pretty good at distilling down complexities and providing a more basic version of the bigger picture, usually with good success. (Given the number of times—about fifty a year—that I am invited to provide lectures to other physicians for continuing medical education and to give public talks on issues related to health, it seems enough people out there think I'm capable of providing some useful education.)

And finally, I believe that as physicians, we should care about the health of everyone and not only the patients who happen to visit us in our offices. On this point there are a number of physicians who agree with me: as a member of the Canadian Association of Physicians for the Environment, I am one of almost five thousand doctors in Canada who consider it important to tackle the broader issues of health, including the environment. (You can learn more about this important and worthwhile organization at www.cape.ca.)

Because of the extremely complex nature of the climate crisis and its potential solutions, I don't believe any one individual has all of the answers. I certainly don't, and for that reason, I'm not going to preach. I endeavour to be as factual as possible, but my opinions will be evident when I think I need to declare them. The truth is, I still wrestle with what I think is the best way for our planet to tackle some of these problems. When trying to envision the solutions to these problems, I think it's relatively easy to imagine how the green future we're striving toward might look; it's something else altogether to visualize how we're going to get ourselves there and how long it's going to take. I believe the path of our transition will be the toughest aspect to predict.

Education is the key. If more people understand these issues, then more people can begin to contribute toward finding solutions. Reading this book will get you on the path to being one of those individuals. My sincerest hope is that you can use the information contained in these pages as a stepping-stone to helping save this planet from ourselves and for our future.

SECTION 1
THE BACKGROUND

Chapter 1

Earth, Its Atmosphere, and Where Carbon Came From

> Space travel has given us
> a new appreciation for the Earth.
> We realize that the Earth is special.
> We've seen it from afar.
> We realize that the Earth is the only
> natural home for man we know of,
> and that we had better protect it.
>
> —James Irwin, American astronaut

Have you seen the most famous photograph ever taken? It's been featured so many times in so many places that it's become instantly recognizable to almost everyone on the planet. Before you purchased this book, there's a high likelihood that you'd seen it already, but now I can guarantee that you've seen it because it's incorporated into this book's cover. It's a spectacular image of Earth taken through a tiny window from *Apollo 17* on December 7, 1972, by crew members Gene Cernan, Ronald Evans, and Jack Schmitt, about five hours into their trip to the moon when the capsule was about 45,000 kilometres or 28,000 miles from Earth. (Credit for the photo is generally shared because no one is sure which astronaut took this particular shot.)

Of all the photos taken of our home planet by the Apollo missions, this one proved to be the most stunning because it offered a full Earth in all her majesty. *Apollo 17* was the only Apollo mission to witness Earth almost completely illuminated by the sun while it was heading to the moon. All of the other missions had only partial views of Earth. (The moon doesn't always have a "full Earth" in the same way that we don't always have a full moon.) It was an early December morning in North America, about 5:30 a.m. EST, so Africa is easily seen during its daytime hours. Since it was only two weeks before the winter solstice, Antarctica is easily spotted at the bottom of the photo because the South Pole points toward the sun at that time of year. The *Apollo 17* crew didn't know that the photo they took would become one of the most reproduced images of all time. They had a job to do, and they did it. But we're forever in their debt because 99 percent of pictures showing the planet Earth are this very image. Nothing else to that point in our civilization's history had ever done a better job of making us realize how special our home is than that single photo.

Earth's atmosphere visible from space.

Since the Apollo missions ended and humans stopped visiting the moon, our astronauts have stayed much closer to home, generally hugging the planet rather than leaving it, at least relative to the days of Neil Armstrong, Buzz Aldrin, and Alan Shepherd. Since those early days of space flight, the vast majority of photos taken of Earth have been from orbiting satellites, from astronauts on shuttles, or from the International Space Station. These modern-day images of our home planet are taken much closer to Earth and can't capture it in its entirety the way *Apollo 17* did.

The next time you look at one of these more recent pictures, pay particular attention to the thin blue strip surrounding the surface of our planet. You probably haven't noticed it before because it's much easier to spot the swirling white clouds, the bright blue oceans, and the interesting and varied land masses composed of dark green forests and light brown deserts. (It's also possible that the thin blue strip isn't there depending on how the photo has been modified for the publication you're reading, but if it's been reproduced properly, it should be noticeable. In black and white photos, it will appear gray.) That thin blue strip is extremely important to you and me: it's Earth's atmosphere, a truly remarkable component of our planet.

Relatively speaking, it's a rather small part of the whole picture, so it's understandable that it gets so little attention. It would be equivalent in scope to a layer of varnish on a basketball. Despite it being such a tiny part of our planet as a whole, we have to appreciate that we live out our entire lives within this thin blue strip. Just think about what happens here: it's all of our weather, from every summer breeze you've ever enjoyed to the most powerful thunderstorms, hurricanes, and tornados; it's our means of travel, not just for every sailboat that uses the winds to set sail to, or for every plane that has flown through it, but also for every vehicle on the road that requires the combustion of gasoline or diesel in the presence of oxygen in order for the motor to work.

Some of the beautiful vistas we take for granted are due to the presence of this atmosphere as well: every sunrise and sunset, every rainbow, every twinkling star that young children wish on, and the northern lights—also known as the aurora borealis and generally seen in latitudes closer to the North Pole—wouldn't exist were it not for this thin blue strip. (There's an equivalent phenomenon called the aurora australis, or the southern lights, located at the South Pole, but with less of the global population living near that part of the world, it isn't as well known.)

And most importantly, of course, life itself depends on this thin blue strip: we breathe in oxygen and breathe out carbon dioxide. Fortunately for us, plants largely reverse this process by using carbon dioxide and water in the presence of sunlight to make oxygen and carbohydrates for their own nutrition, a process called *photosynthesis*. In fact, other than the very few select humans (and even fewer nonhumans, including some chimps and dogs) who have had the privilege of being astronauts, everyone alive today and everyone who has ever lived in the many millennia before us has been born, lived out his or her life, and died within Earth's atmosphere without ever escaping its bonds. Because it is such an intrinsic and routine part of our daily lives, most of us simply take it for granted without ever appreciating how important, precious, and truly vital it is.

The Story of the Elements

To understand Earth's atmosphere, along with the greenhouse gases such as carbon dioxide that are affecting it, an appreciation of carbon itself is useful. Most people know something about carbon—for example, that it's found in coal, pencil lead, and diamonds. Most people also know something about carbon dioxide—that we exhale it, that it's in our soft drinks, and that in solid form it's known as dry ice. But few know where carbon originally came from and, more particularly, how it found its way into fossil fuels deep underground. For the necessary background, we need to go through a bit of Earth's history, but not what we normally think of as history, the kind that deals with the story of human civilizations in centuries past. I'm referring to the history of Earth itself and how it came to be a rocky planet orbiting a star we call the sun, just one of eight planets in our particular solar system located within the Milky Way galaxy. (Since 2006 we have had only eight planets; that was the year Pluto was relegated to its new official status as a "dwarf planet.") To review this background history first requires an understanding of some basic chemistry and, most importantly, an understanding of the elements themselves.

Elements are the most fundamental building blocks of matter. (For those readers who have a deeper understanding of physics, I acknowledge that there are even more elementary particles known as quarks with such interesting names as "strange" and "charm," but that degree of detail is not required here. My apologies to any particle physicists out there who may be reading this.) The elements are familiar parts of everyday life, and you're already well acquainted with many of them. These include elements such as gold, silver, oxygen, nitrogen, lead, nickel, aluminum, sulfur, iron, sodium, calcium, potassium, neon, and helium, and that's just to name a few. There are other elements you have heard about and perhaps even have had some direct interaction with on occasion but aren't exposed to as frequently; these include barium, phosphorus, silicon, iodine, chlorine, uranium, and plutonium. Many elements are extremely rare and are manufactured only in physics labs with names that sound like they're taken right out of science-fiction stories, such as einsteinium, ununoctium, and seaborgium. (Alas, unobtainium, which was so highly sought after in the movie *Avatar*, is not a known element—yet!)

As of 2011, 118 elements have been identified in total. The most recently discovered ones do not occur naturally on Earth, however, but rather are created in particle accelerators—tools that physicists use to smash two lighter elements together in the hopes that they'll combine, albeit briefly, making a heavier element in the process. These manufactured elements survive for only fractions of a second before decaying back into other lighter elements through radioactive decay.

Regardless of whether they occur naturally or are manufactured by physicists, all elements are defined by how many *protons* they have in their centre core, or *nucleus*. The number of protons in the nucleus is referred to as the element's *atomic number*. Each proton has a positive charge, and to make the atom neutral overall, *electrons* surround the nucleus. Electrons have a negative charge, which balances out the positive charge of the protons. There are *neutrons* located in the nucleus as well, which are helpful for stabilizing it. However, neutrons have a neutral charge and don't affect the atomic number; that distinction rests with the proton alone.

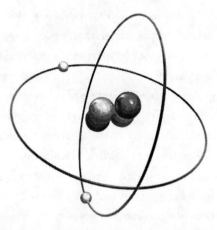

A helium atom consisting of two protons and two neutrons surrounded by two orbiting electrons.

These tiny combinations of protons and neutrons concentrated in a nucleus surrounded by orbiting electrons are called *atoms*, and they are the smallest bits of matter an element can be and still be an element. They are frequently depicted as miniature solar systems in illustrations, although the reality of what they look like is a lot more complex and bizarre. For our purposes, all you need to understand is that each element has a unique number of protons and with it a matching number of electrons, if it is to be electrically neutral, which is the natural state for many of the elements. Hydrogen is the simplest element possible because it has a total proton count of one, orbited by one electron. Helium has two of each; carbon has six, oxygen eight, and sodium eleven. Some of our precious metals are much higher in number, with silver having 47 and gold 79. If you know the name of the element, then you can correctly identify its atomic number and vice versa.

Although a chart could simply consist of just two columns, one with the atomic number of each element from 1 to 118, and the other with its corresponding name, a much more practical and useful chart is used today. Scientists place the elements in an organized fashion in something known as the *periodic table*, the creation of which is credited to the Russian scientist Dmitri Mendeleev. He presented his ideas in 1869, and although other scientists before him had proposed similar concepts for organizing the elements, credit for the periodic table generally goes to Mendeleev. The periodic table is useful because there are relationships among certain elements, a concept known as *periodicity*, with groups of elements having certain common features—most of them placed in the same column—but each element is unique. In fact, Mendeleev was able to predict the characteristics of some elements that had yet to be discovered when he first published his periodic table, and when those elements were ultimately found, and his predictions were confirmed to be remarkably accurate, his place in history was ensured.

The periodic table.

WHERE THE ELEMENTS CAME FROM

We now switch from chemistry to cosmology, the study of the universe. If you pick up any current book or magazine on astronomy, you'll read that it is generally accepted among the scientific community that all of the matter in the universe was created in something known as the *big bang* around 13.7 billion years ago (Hawking and Mlodinow, 50). This was a massive explosion on such a colossal scale that it's not only hard to imagine but also difficult to explain scientifically. (If creationists want to believe in the divine intervention of a Supreme Being, I'd suggest that the big bang is the best place for them to consider looking.) The most abundant element produced in the big bang, subsequently the most abundant element in the universe, was hydrogen. This isn't hard to imagine because it's the simplest element possible with its solitary proton. As a result, it has the lightest weight of any element in the periodic table on an atom-by-atom basis. (If you have an equal number of atoms of any particular element, those with lower atomic numbers will weigh less than those with higher atomic numbers for the simple reason that they have fewer protons and neutrons contributing to their mass. These protons and neutrons may be very small, but they still have mass nonetheless. Electrons, on the contrary, have almost no mass at all.)

Despite such a tiny mass, hydrogen still exerts a small attraction to other atoms, and since hydrogen was so plentiful in the early universe, odds are that if a hydrogen atom was going to be attracted to something else, it was going to be another hydrogen atom. In fact, the hydrogen we know in its gaseous state is actually two hydrogen atoms bonded together, known as H_2. (Once atoms bond with other atoms, they become *molecules*. A hydrogen molecule, therefore, is formed out of two hydrogen atoms.) If you get enough hydrogen molecules in close proximity to each other, they start to coalesce together under the influence of their mutual attraction and can form a ball of hydrogen gas. Although the universe is vast—larger than we can easily imagine—there is still an attraction among all of the atoms and molecules within it, and given enough time, say a few billion years, a large number of hydrogen molecules will attract together to collect into a mass that is significant enough to exert a gravitational force. If you have plenty of hydrogen (a whole lot in fact, something in the unfathomably large ballpark of around 1×10^{30} or 1,000,000,000,000,000,000,000,000,000,000 kg), then you'll have enough to lead to the formation of a star.

That incredibly large number in the last paragraph is just about the mass of our sun, which is a rather average and ordinary star, at least as far as stars go. It may seem hard to appreciate that enough hydrogen can even reach the critical mass to collapse and form into a star in the first place, but when you have billions of years to work with, it turns out it's quite a common occurrence. Case in point: there are probably at least 200 billion stars in a typical galaxy, such as our own Milky Way, and likely close to 200 billion galaxies in the observable universe, making for a grand total in the neighbourhood of 40,000,000,000,000,000,000,000 stars in our entire universe. (I'd say that makes stars a rather common occurrence, even if it is a long way to our closest star neighbour; Proxima Centauri is about 4.2 light-years away, which means that, traveling at the speed of light, it would take 4.2 years to get there. That's about 270,000 times the distance between the Earth and the sun. Don't go booking any vacations just yet.)

So the sun is not much more than a big ball of hydrogen, and yet it gives us all the energy we need to survive. It's not burning its hydrogen by means of combustion, like the Hindenburg explosion in Lakehurst, New Jersey, back in 1937, however. Combustion takes place in the presence of oxygen and isn't so easy to accomplish in the vacuum of space, or in the centre of a star. This is a good thing too because if our sun burned its hydrogen by combustion, it would have burned out billions of years ago, and you wouldn't be reading this now. A star's fuel is used through an entirely different process called *nuclear fusion*, or simply fusion for short.

A star is a massive object, even an average star such as our sun. As an example, everything in our solar system, including every planet, comet, asteroid, and dwarf planet, could easily fit within the volume of our sun and still leave plenty of room to spare. The density and the effects of a star's own gravity at its very centre is extremely strong, making for a massive amount of pressure that you and I can't possibly imagine. In fact, the pressure at the centre of a star is so extreme that hydrogen doesn't even exist as H_2 but rather as individual hydrogen atoms, stripped of their electrons and unable to form into molecules. It's more like an extremely dense soup of protons and electrons mixed together. Because of that intense pressure, however, they are particularly close to other hydrogen atoms in the same situation. (Cramming into the subway at rush hour wouldn't even begin to give you an idea, but it's a useful analogy to consider.) The pressure combined with the unimaginably high temperatures is enough to actually squeeze two individual hydrogen atoms together and combine them into one larger nucleus. Since there's one proton in each hydrogen atom, the combination results in a nucleus of two protons. An atomic number of 2 corresponds with the element helium, the same stuff that keeps your party balloons afloat and gives you the voice of a Bee Gee for a few moments after you inhale it. (For the record, I don't endorse breathing in anything that might deprive you of consciousness.)

This process is precisely what fuels a star: nuclear fusion combining two hydrogen atoms into one helium atom. It should now seem rather obvious why this process is referred to as fusion: it's simply the fusing of two atomic nuclei together, creating an element with a higher atomic number. (This is in contrast to a different nuclear process called *fission*, where a larger element splits into two smaller elements. Chapter 7 includes more information on fission and how nuclear energy is obtained.)

You might be asking yourself how any energy comes from such a process: why would two hydrogen atoms combining to become one helium atom create any energy? It has to do with a little equation known as $E=mc^2$, made famous by Albert Einstein and first published in a seminal paper he wrote in 1905. I would dare say that almost every one of you reading this book has heard of this, the most famous equation of all time, but few truly understand what it means. In a nutshell, Einstein helped to prove that energy (E) and matter (m) are related and interchangeable, essentially different forms of one another.

This is difficult to comprehend in the daily routine of our lives, but we still experience the effects of this equation on a regular basis. Most importantly, we wouldn't be here if what it describes didn't power the sun. (This same equation explains the significant energy derived from an atomic bomb, but I prefer to think of the good that comes from science rather than some of its more negative applications.) The c in the equation refers to the speed of light, which is about 300 million metres per second, or 186,000 miles per second. It's beyond the scope of this book to explain why the speed of light is part of this equation. (At the end of this chapter, in the section titled "Suggested Reading for Chapter 1," I list two very useful and readable books on the subject if you want to explore this very interesting topic in greater detail.) For the most basic of explanations, let's first look at an important equation taught in high school physics known as the equation for kinetic energy. This equation explains the amount of energy an object has simply as a result of its motion. An object with small mass moving quickly, such as a bullet, and an object with large mass moving slowly, such as a rolling vehicle, may have a comparable amount of kinetic energy; this equation helps to determine precisely how much.

Kinetic Energy (E_k):
$$E_k = \tfrac{1}{2}mv^2$$

In this equation, *m* refers to the object's mass and *v* to its velocity. You can see that the energy of a body in motion is related to the square of its velocity, in other words the velocity multiplied by itself. Thus, every time you double the velocity of a moving object, you increase its energy from that motion by a factor of four times. That's why a car crashing into a brick wall at a high speed results in a lot more damage than it would moving at a slower pace. The mass of the car is the same in both situations, but the marked difference in energy levels is due to the difference in the velocities. Obviously, a heavier mass will move with a greater amount of energy, as the equation also shows, but doubling the mass only doubles the energy: that part of the equation is linear.

Now think of the speed of light as an absolute speed limit that nothing can pass. The amount of energy equal to an object's mass is related to the square of this absolute speed limit. Since the *c* in Einstein's equation is squared, the number becomes 300,000,000 × 300,000,000, for a grand total of 90,000,000,000,000,000. It's easy to see that even a small amount of matter multiplied by this very large number equals an extremely large amount of energy, regardless of what units they are measured in. (The units of energy in the equation happen to be joules, for those of you who are interested.)

This is precisely what happens with the fusion of two hydrogen atoms into one helium atom in the centre of our sun. It's not just a matter of one plus one equals two, with two hydrogen atoms combining to become one helium atom and nothing more. In fact, a minuscule amount of matter is lost in the process and converted into an extremely large amount of energy. Every star in the universe is fueled in precisely this way.

As stars get to the point where they are running out of hydrogen to use for fusion, they can start to use the helium they have been generating and fuse it into even heavier elements. Fusion of helium doesn't produce as much energy as the fusion of hydrogen does—at least compared with the energy invested in getting the atoms to fuse in the first place—but it still allows the star to keep burning. This can lead to the creation of the next few elements down the line in the periodic table, which are lithium, beryllium, boron, and carbon. (You knew carbon was going to come into the picture eventually!) They each have atomic numbers of three, four, five, and six, respectively. If the star is large enough that its slowly diminishing mass still has enough gravitational force to keep the pressure cooker going at its centre, then once the helium is used up in the fusion process, it can move down the line onto the carbon it created.

Once again, fusion using carbon atoms will produce even heavier elements. Fusion with carbon has an even worse rate of return than fusion using helium. As you go down the fusion ladder, each step produces less and less energy relative to the energy required to allow the fusion to take place. Depending on how massive the star is and how fast it goes through its fuel, a point will be reached—usually a few billion years—where the energy created is outweighed by the energy it takes to allow the reaction to occur. This process continues right up to the production of iron, with an atomic number of 26. It turns out it takes more energy to fuse something onto iron than is obtained from the fusion reaction itself, despite the massive amounts of energy created from the process depicted by the equation $E=mc^2$.

The Origin of Elements beyond Iron

So how did anything past iron ever get created then? Where did all of these other elements with higher atomic numbers come from? Once a star has burned up all of its possible sources of fuel through the process of fusion, there are a number of ways it can die. Some just simply burn out and go dark. A white dwarf is an example of a star that just didn't have what it takes to continue past the stage of producing carbon and oxygen, for example. (Interestingly, white dwarfs are still very hot and still give off light; the universe isn't old enough for them to have completely burned out, but they are no longer powered by fusion.)

If a star is massive enough, at least one and a half times as massive as our own sun, then it can experience one of the most dramatic stellar deaths possible in a tremendous explosion known as a *supernova*. In fact, a supernova is so massive an explosion that it temporarily outshines the light of all the other 200 billion stars in its galaxy—combined! A supernova contains more energy than our sun has produced in its entire lifetime. Each galaxy has about five of these every century on average, so they aren't particularly common. The last observed supernova in our own Milky Way galaxy was in 1604, so we're long overdue, but with billions of galaxies out there, we get to see them from time to time when we're really looking, as our astronomers always are.

The cause of a supernova has to do with the collapse of a star that no longer has enough energy generated from fusion to push against the compressive effects of its own gravity. The resulting mega-explosion that results from this collapse, and the heat and pressure associated with it, can lead to all sorts of fusions of elements much heavier than hydrogen, helium, carbon, magnesium, and iron, although it's still the same fusion process described earlier. There are still plenty of elements up to and including iron after such an explosion, but there will also be many more elements heavier than iron, with higher atomic numbers. These include silver (47), tin (50), and gold (79). The element with the highest atomic number that is found naturally on Earth is uranium, at 92, but it is found only in trace quantities, which is fortunate given its ability to be used in nuclear weapons. The supernova energy would create elements beyond uranium as well, but because they are less stable and experience radioactive decay more quickly, none of those are found naturally on Earth any longer.

Once you go past uranium in the present day, the rest of the elements usually have to be manufactured in particle accelerators as described earlier. The higher the atomic number of the element, the more quickly it will tend to decay. (Plutonium, with the atomic number of 94, isn't found naturally and has to be manufactured; it was first made from uranium in 1940. It has, of course, found its way into nuclear weapons laboratories rather than laboratories used purely for the purposes of scientific discovery, unfortunately.)

Eventually, after the remnants of the supernova have settled down—give or take a few more billion years—gravity can take over and start collapsing things back together again. Because the simplest element is so plentiful in the universe, there is still enough hydrogen around to collapse again to form another star at the centre of what will eventually become a solar system. The elements that were blasted away in the supernova's explosion eventually form into smaller masses that orbit that new second-generation star. The supernova blasted the lightest elements farther than the heavier elements that it created, simply because the energy required to move something lighter is less—think back to the equation for kinetic energy. As a result, the outer reaches of this newly forming solar system tend to be planets formed from gases because the lightest elements—namely hydrogen and helium—are gases rather than solids. They still form by a mutual attraction among their atoms and manage to combine into large enough collections to become planets—but not so large to become stars themselves—after billions of years. They're called *gas giants* because they tend to be huge and don't have a rocky surface that you could stand on—if you ever managed to make it there for a visit. In our solar system, the outer planets Jupiter, Saturn, Uranus, and Neptune are all gas giants.

The regions closer to the centre of this new solar system contain more of the heavier elements that were formed in the supernova, so they initially attract each other and coalesce into tiny but solid particles. At first, small grains form that are no larger than the size of dust particles, typically made out of iron, silicon, and other metal compounds. These particles combine together in time, forming into progressively larger sizes. They form small rocks first and then larger boulders, ultimately becoming objects known as *planetesimals*, which are large enough—about one kilometre or a little over half a mile across—to have a substantial gravity. These continue to grow in size, with further particles bombarding them, and eventually they become large enough to become planets. Unlike the outer gas giants, however, these inner planets will eventually become rocky. In our solar system, these planets are Mercury, Venus, Earth, and Mars.

All of these collisions that produce these inner planets create intense heat, so there is a lot of melting and vaporization during the initial stages of the formation of a new solar system. These particular planets are initially hot, molten balls. Around this time in their development, the new star they're orbiting reaches a point where there is enough pressure at its core for it to "turn on." It begins to produce its own energy as heat and light via the same processes of fusion that powered the earlier generation of stars from which this one descended.

As time passes—again we're talking about a few billion years—the planets slowly cool. The denser iron settles in the centre. Ever wonder why our planet has magnetic poles? It's just a big bar magnet with an iron core that is still molten. The lighter elements cool and harden on the surface—in the case of our planet, forming Earth's crust. Think of a pudding skin that forms on the surface as some freshly made pudding cools (for those of you who remember that pudding wasn't always purchased in a ready-made container from a grocery store).

Many of the elements created in the supernova become part of the crust, and most elements up to and including uranium can be found somewhere on

Earth because of this process. They aren't always evenly distributed, which is why there are a lot of diamond mines in South Africa—diamond is a particular form of carbon with a crystalline structure—and significant gold deposits in California as well as the Yukon, which led to the gold rushes of the nineteenth century. These elements aren't as abundant everywhere else as they are in those particular geographical regions, but every element up to uranium can be found in its natural state somewhere on Earth.

Everything in our solar system, from the unused hydrogen in our sun to the gold and silver on Earth, is a remnant from an earlier star's supernova. It's interesting to think that all of the gold rings, silver bracelets, and diamond necklaces that people treasure are made of materials formed in the massive death throes of a star that died billions of years ago. Consider it phase 2 of the universe's history. Phase 1 occurred with the first stars that were born from the earliest hydrogen available after the big bang; these stars lived out their lives over billions of years and then died in fiery supernovae. Phase 2 began when new stars and their solar systems formed using the residual materials left behind from the earlier supernovae. Astronomers have used a variety of complicated techniques to determine that our universe is about 13.7 billion years old, but our own solar system is closer to about 5 billion years old, a relative newcomer on the cosmic block. As Carl Sagan used to so eloquently state, we are all "star stuff." Everything we know in our entire world, along with all of the other planets, moons, asteroids, and comets that are part of our cosmic neighbourhood, came from a massive explosion that happened billions of years ago.

THE DISTRIBUTION OF CARBON ON EARTH

Perhaps you're wondering what exactly this has to do with global warming or even the atmosphere. From the processes just described, you can now understand where all the elements on Earth have come from. With very few exceptions—namely those elements beyond uranium that are created by particle accelerators for brief moments, or any elements that have experienced radioactive decay naturally—the vast majority of atoms that exist on Earth today have existed as atoms with the same atomic number and identity since our solar system was initially formed. The absolute amount of any one element is generally fixed. These elements can mix and match in different ways to form various molecules, so the oxygen atoms on Earth might be part of the oxygen you're breathing right now, soon to be part of a carbon dioxide molecule that you'll exhale. Oxygen atoms also could just as easily be part of the alcohol molecules in the glass of wine you had with dinner, or the sugar molecules in your dessert, or even part of some calcium carbonate in a shell in your fish tank. But no matter how you look at it, these oxygen atoms are still the same ones they've always been for the billions of years that they've been present on our planet, just mixed around in different molecules at different times.

The same applies to all elements, not just oxygen. The one we're going to discuss most, the one on which you will be more of an expert by the end of this book, is carbon. Just as with all the other elements, for all practical purposes every carbon atom on Earth has been here since the very beginning of our solar system about 5 billion years ago. They too are remnants of that same supernova that occurred earlier still. These atoms might be carbon in its purest form, such as coal used for fuel or graphite used in pencils (pencil "lead" is a misnomer). Or perhaps these carbon atoms are organized in carbon's rarest natural form on Earth as diamonds. (Diamonds' crystalline structure with a complex network of carbon atoms linked together is what gives diamonds their properties of reflecting light in such amazing and beautiful ways.)

Of course, carbon atoms aren't necessarily on their own but might be combined with other atoms, such as oxygen in carbon dioxide molecules like those you're exhaling right now, or as part of the sugar molecules in your dessert, or in the calcium carbonate in a shell in your fish tank. (The redundancy in what you've just read is intentional; you'll soon learn that hydrogen, oxygen, and carbon have special relationships that are critical for life to occur. Hydrogen, oxygen, and carbon are all quite common after a supernova, so there are plenty of these atoms around. That's one of the reasons they repeatedly bond with each other.)

The vast majority of carbon atoms on Earth—about 99.9 percent—are locked up inside the rock situated within our planet's crust, estimated by geologists to be about 50,000,000,000,000,000 tons' worth (Mathez, 59). (To make that a little simpler to read for the purposes of comparison, I'll refer to a billion tons as a *gigaton*. Thus, the amount of carbon in Earth's rock is 50 million gigatons, which is a lot.) There is plenty of carbon in Earth's oceans as well, partly absorbed as carbon dioxide in the water, partly in the calcium carbonate shells of crustaceans, partly in the plants and animals teeming throughout the waters surrounding our planet, and partly in the sediments on the ocean floor. The amount of carbon in our oceans amounts to about 39,000 gigatons in total. On land there is carbon in the soil (1,580 gigatons) and in all of the plant life as well (610 gigatons). And the atmosphere, that thin blue strip referred to at the beginning of this chapter, contains about 750 gigatons of carbon in the form of carbon dioxide.

As described previously, atoms can exist on Earth in various ways, whether they are hydrogen, oxygen, carbon, or any of the other numerous elements found on our planet. They can be with other atoms of the same element, such as carbon in coal or diamonds, but they can also mix and match in various ways to make up a variety of different molecules. It's also important to realize that these atoms aren't static; they aren't stuck in the same spot for all of eternity. Instead they circulate from one source to another. Chapter 2 explores how carbon makes its way through all of these various sources in an important and completely natural process known as the carbon cycle.

Key Concepts of Chapter 1

- Our atmosphere is a significant part of our planet, with all aspects of life dependent on it.
- All of the elements on Earth and in our solar system were created from an earlier supernova through the process of nuclear fusion.
- Carbon is present all over our planet in various forms, often combined with other atoms as part of more complex molecules.
- Carbon, hydrogen, and oxygen are three of the most important elements necessary for life.

Suggested Reading for Chapter 1

Bodanis, David. *E=mc²: A Biography of the World's Most Famous Equation.* New York, NY: Walker, 2000.

Bryson, Bill. *A Short History of Nearly Everything, Special Illustrated Edition.* Toronto, ON: Doubleday Canada, 2005.

Cox, Brian, and Jeff Forshaw. *Why Does E=mc2?* Cambridge, MA: Da Capo Press, 2009.

Hawking, Stephen, and Leonard Mlodinow. *The Grand Design.* New York, NY: Bantam, 2010.

Kean, Sam. *The Disappearing Spoon: And Other True Tales of Madness, Love, and the History of the World from the Periodic Table of the Elements.* New York, NY: Little, Brown, 2010.

Mathez, Edmond A. *Climate Change: The Science of Global Warming and Our Energy Future.* New York, NY: Columbia University Press, 2009.

Chapter 2

The Carbon Cycle: How Mother Nature Circulates Carbon and Generates Fossil Fuels

> I would feel more optimistic
> about a bright future for man
> if he spent less time proving
> that he can outwit Nature
> and more time
> tasting her sweetness
> and respecting her seniority.
> —E. B. White

Earth is a very dynamic place, and everything on it moves given enough time. We sometimes forget this because we tend to think in terms of a human lifetime, but in geological terms, over a span of millions or even billions of years, things move. The carbon on our planet is no exception. To increase understanding of how carbon moves around on Earth naturally—something known as the *carbon cycle*, referred to at the end of the last chapter—we will look at some examples that are easier to relate to, first by following the travels of a water molecule and then by following an atom of oxygen. After that, the travels of carbon atoms will be easier to understand.

Earth's surface is mostly water; think about that fact for a moment. Water covers about 75 percent of our planet, with land making up only 25 percent. The amount of water on Earth is so enormous that the Pacific Ocean alone is larger than all of the land masses on our planet combined. Because it's so plentiful, our planet looks blue from space in any colour photos that have been taken of it. That's why it's often referred to as the big blue marble.

Water is a vital part of life, and every plant and animal on land and in the sea depends on it for survival. Based on data sent from spacecraft that have been sent out to investigate other planets, Earth is the only one in our solar system where water exists in any abundance. Our planet is also uniquely placed at just the right distance from the sun—about 150 million kilometres or 93 million miles away—to allow water to exist in all three states of matter naturally, with ice closer to the North and South Poles, liquid water in the warmer climates, and the gaseous state of water occurring throughout the atmosphere (the clouds are proof of the latter). Our planet isn't too hot or too cold; it's been just right, as Goldilocks would say, for life to exist.

Of course, water isn't important only for the life cycles of the plants and animals on Earth. It also plays a significant role in our planet's weather. A summer shower wouldn't exist without it. Neither would a thunderstorm, tornado, or hurricane. Yet one thing that is easily appreciated is that water on Earth is constantly on the move. Water molecules lead a dynamic existence. As a case in point, let's follow a water molecule and see where it might take us.

The Travels of a Water Molecule

To trace the path of a molecule of water, we'll first start in a cloud. Our water molecule might linger there for a few days, but it won't take long before it falls to Earth's surface as part of a rain shower. Maybe it will land in an ocean (a good bet given the percentage of our planet covered by oceans), or possibly somewhere on land, such as the garden in your backyard. Obviously, some land masses such as deserts have less rainfall and would be less likely.

For the sake of our example, we'll imagine that it falls into a freshwater lake somewhere inland—perhaps the one you like to fish in. Once it's made its way into that lake, it will remain there for a while too, but eventually, it will move on simply because of the natural currents present in any body of water. In our hypothetical case, it happens to flow into a river that comes off as a tributary of the lake, and this river supplies some much-needed irrigation to several farmers' fields nearby. That water molecule can then irrigate the soil so that it can be taken up by the crops that will grow into some of the fruits and vegetables we eat every day, just like the produce you buy at your local farmers' market or grocery store. Perhaps that water molecule becomes part of a vegetable, such as a potato, mushroom, or carrot. It might just as easily become part of a fruit, such as an apple, orange, or banana. But one thing is for certain: once that fruit or vegetable is eaten, that water molecule becomes part of the critter that ate it. In our story, that critter is you!

There are many places where water exists inside of you because it's an intrinsic part of almost every one of the 10 trillion cells in your body. Sixty percent of you is made up of water, in fact. (My sons Matthew and Jamie recently discovered at a Toronto Science Centre exhibit that their bodies are made up of about twenty-two and twenty litres of water, respectively, both between five and six gallons; they thought that was pretty cool.) Any excess water your body doesn't need is filtered out in your kidneys and disposed of in urine, but it never takes you very long to become thirsty again. This is because your body requires a fairly precise amount of water, not too much and not too little, but just right.

Your water molecule will eventually circulate throughout much of your body. After you swallow it as part of the fruit or vegetable you ate, it will find its way down your esophagus and into your stomach. It will then travel through your intestines and eventually be absorbed, along with the nutrients from the food you ate, into your bloodstream. From there it can travel almost anywhere. For our purposes we'll arbitrarily have it make its way into one of your sweat glands. When you get a vigorous physical workout while playing some recreational sports, your water molecule may eventually be secreted as perspiration. Some of your sweat will evaporate on your skin's surface via heat energy from your skin, cooling you off in the process—the very reason you perspire in the first place.

The molecule of water from your perspiration evaporates and travels back into the atmosphere, possibly to join another cloud and start the whole cycle all over again. (It turns out that a molecule of water takes about nine days on average to circulate through the body before it finds its way out for some other destination. That particular fact has been confirmed by the use of a form of water that can be easily traced. Radioactive water exists when the hydrogen in the water molecule has a neutron in addition to its solitary proton and can be detected with specialized equipment such as a Geiger counter. This type of water is known as heavy water because it has a greater mass than the usual variety of water due to the additional neutron. About nine days after we ingest heavy water, it turns out most of it has left us.)

And so it goes, as water circulates throughout the planet in various ways: it moves back and forth between land and air, travels through plants and animals, moves into oceans and rivers, and is absorbed into soil and evaporated into the atmosphere. Sometimes water molecules freeze into ice if they find their way to latitudes closer to the North or South Poles. If they end up in the more moderate latitudes, such as the northern United States or southern Canada, then the ice will be seasonal, eventually melting back into liquid water the following spring. But if the water molecules are even closer to one of the poles, they may become part of the permanent polar ice where they can remain unchanged for centuries. No matter where they end up, though, the water molecules on our planet were elsewhere before; given enough time, they will end up somewhere else again. They are always on the move.

The Travels of an Oxygen Atom

Other molecules easily travel through various regions in our planet as well, just as water molecules do. Gases are the lightest and therefore the most mobile, so carbon dioxide, oxygen, and nitrogen have the easiest time getting around. Liquids move quite easily as well, as our nomadic water molecule demonstrated. But even molecules in solids get around. (Just think of the sand in your shoes that you brought home from the beach. The wind-swept sand dunes in deserts demonstrate that solid substances such as sand move around naturally as well and do not have to rely solely on people power.) Just as molecules travel about, it's important to understand that the individual atoms within molecules also get around because the chemical bonds that hold molecules together can be broken when they are exposed to some form of energy, such as heat. The breaking up of chemical bonds is what defines a chemical reaction; it frees up the individual atoms within a molecule, allowing them to join up with other atoms to create new molecules. In other words, a chemical reaction occurs when the atoms within molecules break apart and rearrange themselves into different combinations.

Let's take the example of the oxygen atom within our water molecule. Water molecules are made up of two hydrogen atoms and one oxygen atom. The term H_2O is familiar to everyone, even those who know nothing about chemistry. Instead of exclusively remaining part of a water molecule, maybe this particular oxygen atom can be used by a plant in *photosynthesis*. This process is used by plants all over the world, including algae, which are particularly plentiful in the 75 percent of our planet covered in water.

The *chlorophyll* in these plants that gives them their distinctive green colour is where photosynthesis occurs. Think of chlorophyll as a factory for making *carbohydrates*, which are molecules made up of carbon, hydrogen, and oxygen atoms. These include the energy-containing carbohydrates, such as sugars, and the structural carbohydrates, such as cellulose, that help to give plants their structure and keep them upright. (The trunk of the mightiest oak and the stem of the sweetest rose both contain cellulose, giving them the ability to maintain their distinctive shapes.) The chlorophyll in plants can take molecules of water (H_2O) and molecules of carbon dioxide (CO_2) and combine them together to manufacture carbohydrates out of them. In photosynthesis, six carbon dioxide molecules and six water molecules become one glucose molecule ($C_6H_{12}O_6$).

Photosynthesis:
$$6\ CO_2 + 6\ H_2O + \text{energy} \rightarrow C_6H_{12}O_6 + 6\ O_2$$

You'll notice that there are eighteen oxygen atoms on both sides of the equation: originally, twelve were in carbon dioxide, and six were in water, but after the reaction was complete, there were six in the molecule of glucose and twelve in the six oxygen molecules, two in each. Likewise, you can follow the equation to see that six carbon atoms in carbon dioxide became six carbon atoms in one glucose molecule, and the twelve hydrogen atoms in the water molecules all became part of the one glucose molecule. Everything has to equal out on both sides of an equation that describes a chemical reaction. This is a very important principle known as *conservation of mass* and applies to every chemical reaction. (This is in distinction to nuclear reactions such as fusion, described in the last chapter, where a small amount of matter is converted into energy. In chemical reactions, all of the mass is always accounted for on both sides of the equation.)

And just where does the energy for photosynthesis to drive these reactions and make them possible come from? Obviously, plants don't have fossil fuels at their disposal to use as an energy source like we humans do. The energy for the process of photosynthesis comes from sunlight. "Photo" means "light." (Think of photograph and telephoto lens, and the connection to light will be evident.) Plants have been using solar energy for billions of years to store energy in the form of these various carbohydrates. The by-product of this process is even more interesting because the "waste material," if you will, is oxygen.

So all of a sudden, the oxygen atoms in our water molecules on the left side of the equation find their way into glucose and oxygen molecules on the right side of the equation. If a molecule of glucose combines with a fructose molecule to become a sucrose molecule, it might be used in baking, be used to sweeten coffee or tea, or otherwise ultimately be eaten by a critter, again perhaps you! Once consumed, it becomes part of the energy stores we use every day to fuel our bodies.

And how do we burn the calories stored in sugar? Simply by reversing the equation for photosynthesis. We burn glucose in the presence of oxygen. This is why we have to breathe oxygen into our lungs in the first place, a process known as respiration. We produce carbon dioxide when we subsequently exhale. We also produce water from the use of glucose for energy, and that simply adds to the body's water stores. Thus, the oxygen atom in a water molecule can easily be converted into an oxygen atom in a glucose molecule via photosynthesis, only to be burned as fuel and converted back into part of a different water molecule than the one it started in.

Aerobic Respiration:
$$C_6H_{12}O_6 + 6\,O_2 \rightarrow 6\,CO_2 + 6\,H_2O + \text{energy}$$

Notice that this equation is an exact reversal of the equation describing photosynthesis. I point all of this out because it illustrates how Earth is a dynamic place with molecules constantly on the move from one source to another, and atoms within those molecules such as oxygen are constantly rearranging themselves. The equations for photosynthesis and *aerobic* respiration are what allow life itself to exist on our planet.

Although it's not as intuitive to think about, carbon atoms do the very same thing as oxygen atoms. Carbon atoms may be in the atmosphere in the form of carbon dioxide, in the rocky crust of our planet (carbonate rocks include limestone and marble), or in the various forms of life on land or in water. They may be stored in glucose molecules as a form of energy; in more complex molecules such as cellulose, giving tree trunks their strength; or in the calcium carbonate shells of animals such as lobsters. Human beings couldn't exist without carbon. It's a vital component of life, and it's just as important to us as water is.

To understand the issue of carbon dioxide emissions and how they can impact our climate requires an understanding of how carbon naturally moves about throughout our planet in the carbon cycle. (Just so we're clear, by "naturally," I mean what happens to carbon without the influence of human beings. As will be addressed later, we've had a significant impact on the carbon cycle as well.) This has been going on for billions of years, but because the naturally occurring transitions of carbon atoms from one source to another are often much slower than our water molecule and oxygen atom examples, this isn't as easy to appreciate. The movement of carbon within the planet is often on a much longer geological time scale, in some cases taking millions of years to move from one area to another. Aside from that, there is a lot of similarity between the movement of carbon within the carbon cycle and the travels of water molecules and oxygen atoms.

THE CARBON CYCLE

Let's now tackle how carbon has been moving around Earth since before human activity had any impact. (Our impact will be addressed in section 2.) It was already mentioned in chapter 1 that Earth first formed into a planet around 5 billion years ago from the remnants of a much earlier supernova, which provided it with all of the heavier elements it still has today. Most of the carbon was locked into the Earth itself within the rocky crust along with many of the other lighter elements—with the greater concentration of heavier elements, such as iron, even deeper inside the planet's core. However, the infant Earth was an extremely dynamic, active, and inhospitable environment, incapable of supporting life as we know it today. There were dramatic earthquakes from the unstable crust, bombardments of the surface by countless objects, such as comets and asteroids, still whizzing around the early solar system, and significant volcanic activity, releasing much of the energy contained deep inside the planet. Earthquakes, comet and asteroid impacts, and volcanic eruptions all provided more than enough energy to allow carbon and oxygen to combine together. Through these early chemical reactions then, much of the carbon within the Earth's crust was being released into the atmosphere in the form of carbon dioxide.

This gradually led to very high levels of carbon dioxide in the early atmosphere, perhaps around 1,000 parts per million (ppm). To put that number into some perspective, during most of the last 650,000 years, the concentration of carbon dioxide in our atmosphere has never been higher than 300 ppm, usually more in the range of 260–280 ppm. (Just how that was determined is explained in chapter 4, but if anyone ever points out to you that there was a time in Earth's past when carbon dioxide levels were much higher naturally than they are today, they're telling you the truth. However, this was so early in Earth's history that it doesn't really compare to the present day. We couldn't have survived so easily back then—if at all—with such high levels of carbon dioxide.)

It wasn't until around 3.5 billion years ago that the first life forms came into the picture. The water so necessary for life is believed to have come mostly from the many comets and asteroids that bombarded our young planet's surface. Comets are really large balls full of ice mixed with remnants of the early materials that led to the formation of Earth in the first place. It's easiest to think of them as dirty snowballs. Asteroids are more like big rocks that never had the chance to join with a larger body to become part of a planet. With billions of years having passed since Earth first formed, most of the stray comets and asteroids in our solar system have already been cleared away by the strong gravitational fields associated with the massive gas giants in the outer solar system, sparing Earth from as many collisions today as in the earlier stages of our planet's history.

Many comets and asteroids still exist, of course. Every few years, a comet makes the headlines because it's visible in our night sky, and there is also a large asteroid belt located between Mars and Jupiter that still causes angst for some who are worried that a stray rock will someday collide with Earth and cause our planet significant destruction, perhaps even destroying life in the process. (It has happened before: it's believed that 65 millions years ago, an asteroid impacting near the Yucatan peninsula in modern-day Mexico led to the widespread extinction of the dinosaurs. Hopefully, it won't happen again for a really long time, but it most likely will eventually—unless our technology is advanced enough at that point to identify potential collisions and somehow prevent them.)

Although it is believed that early impacts from comets and asteroids provided most of the water on Earth, and that various necessary inorganic components, such as ammonia, carbon dioxide, sulfur dioxide, and methane, were already present from Earth's formation, how life began on Earth still is not known. All of these molecules can rearrange themselves into the organic molecules necessary for life; these include such compounds as the amino acids found in proteins and the fatty lipid layers in cell walls that give them their shape. The energy required to allow these complex molecules to form out of the simpler molecules was easily provided by sources such as the electricity available in lightning. (The process of taking inorganic compounds and converting them into organic ones has actually been reproduced in a famous experiment first performed in 1952, known as the Miller-Urey experiment after the two scientists who published it in the journal *Science* in May 1953.) What we don't know is how the primordial soup of various inorganic and organic molecules could combine into some form of organism that could grow, metabolize material to provide itself some energy, and reproduce. These are some of the fundamental requirements of a living organism. Since that isn't known, there isn't much point in offering conjecture at this point, and it isn't germane to this part of the story of carbon. Suffice it to say, it happened.

The first life forms were single-celled organisms known as *prokaryotes*, and they were pretty simple as far as life forms go. However, prokaryotes have proven themselves to be a hardy type of organism, thriving for billions of years. As it happens, we still live with them every day of our lives. Sometimes these simple organisms can take humans out of commission with a minor illness, a major illness, or even death: all bacteria are prokaryotes, from the streptococcus that causes a sore throat to the bacillus that causes a urinary tract infection, to the staphylococcus that causes toxic shock, a highly fatal infection.

Although prokaryotes evolve—bacterial resistance to antibiotics is proof of that—they exist largely unchanged compared to their earliest versions that first came into existence 3.5 billion years ago. They also provided an evolutionary stepping-stone for more complex cells called *eukaryotes*. Eukaryotes have a number of features that distinguish them from the simpler prokaryotes, but the most important one is that they have a central nucleus containing their genetic material. (*Nucleus* comes from the Latin word for "core," explaining why the word is used for both the central portion of an atom and the central body in a cell.) They also have more complex methods of cell reproduction. Prokaryotes reproduce asexually by simply splitting their material in half and giving equal amounts to each daughter cell. This works well when something needs to reproduce quickly—again think of how quickly a bacterial infection can take over a previously healthy organism—but eliminates the genetic variation that can be so helpful to evolution. Eukaryotes divide their cells using processes known as *mitosis* and *meiosis*.

Mitosis is the process where a cell reproduces itself into an identical copy—important for you and me to replace the many cells in our bodies that continuously die off, including blood cells, skin cells, and the lining of our digestive tract. It's more complicated than the asexual reproduction used by prokaryotes because it involves duplicating all of the genetic material found in the nucleus first so that both daughter cells have exact copies of the parent cell. Meiosis, on the other hand, is where a cell's genetic material is divided down the middle, with half going into one cell and half in another. Once these cells encounter another cell that was also formed by meiosis, they can combine into a brand-new cell that will be slightly different yet will still share similarities to the two parent cells from which it was formed. If you think this sounds similar to how we procreate, you're exactly right: meiosis is required for sexual reproduction.

Meiosis grants eukaryotic cells the advantage of sharing genetic material and allowing variations to occur from one generation to the next. Every plant, animal, and fungus on Earth is a eukaryote, and although eukaryotes can't reproduce as quickly as bacteria, any variations or mutations that lead to some improvement have a survival advantage and are favoured to continue on in the process, sharing their genetic material, with the associated improvements being passed on. Most genetic mutations are harmful rather than helpful, and so they aren't always as likely to continue as a survival advantage, but every once in a while, mutations make a positive difference. This is essentially the theory of evolution and natural selection, first published by Charles Darwin in 1859.

For now we're going to focus on the eukaryotes that became plants because they played a significant part in affecting the composition of the Earth's atmosphere early on and continue to do so today. It was plant life that, using photosynthesis, helped reduce the early carbon dioxide levels that were so much higher than they are today, thereby increasing the amount of oxygen toward the levels we enjoy presently. Despite the obvious advantages to animals who depend on oxygen for respiration—animals such as us in fact—plants aren't exactly trying to be good Samaritans in creating so much oxygen. Instead, they do what they do for the sole purpose of their own survival. (A rose doesn't smell nice and look pretty for our enjoyment. Rather, it evolved to have a fragrance and an appearance that would attract insects, which would then carry away the rose's nectar, subsequently taking the pollen along for the ride from one rose to another, helping to share the genetic material and allowing future roses to grow.)

Plants need energy to survive; every life form does, and the energy we get as human beings ultimately comes from plants. Unlike human beings who can hunt for food, drive to the nearest grocery store, or grow food in a backyard vegetable garden, plants lack the mobility to procure the food that will provide them with energy. Being fixed in one location, most plants need to make their own food, just like they need insects to get their genetic material passed on from one plant to another. To be fair, there are some plant species, such as the Venus flytrap, that wait patiently for an unsuspecting insect to stumble into the plant's jaw-like leaves, which close on the helpless victim and release enzymes that allow the plant to digest the doomed bug. (In that case, the food comes to the plant—and it didn't even need to phone in for delivery—but that's the exception rather than the rule.) The vast majority of plants, however, solve the problem of not being able to hunt for food by making their own.

Photosynthesis is literally the first form of solar energy used on Earth. As described in the last chapter, the process of photosynthesis not only produces carbohydrates that plants—and ultimately the animals who eat those plants—use as food; photosynthesis also produces oxygen as a by-product. To reiterate, plants use a process where carbon dioxide plus water plus energy from sunlight produces glucose as their primary source of food plus oxygen, as the photosynthesis equation demonstrated earlier. It isn't hard to see now that the early levels of carbon dioxide that were so much higher than they are today would start to decline as plant life began to absorb it right out of the atmosphere to produce their own food energy by manufacturing carbohydrates. Likewise, it isn't difficult to see how the oxygen levels would gradually increase in the atmosphere as a direct result of the same chemical process. It took billions of years for the levels to get close to where they are now. At present, oxygen makes up about 21 percent of the composition of Earth's atmosphere, and carbon dioxide is only a very small amount, for many millennia at levels between 260 and 280 ppm.

How Carbon Gets Around on Land

Similar to our earlier tours where we traced the paths of a water molecule and then an oxygen atom, we'll now follow the path of an atom of carbon on land. Carbon atoms can exist as part of the carbonate rocks in the Earth's crust, only to be released into the atmosphere as part of carbon dioxide molecules through volcanic activity. (As much as 40 percent of the carbon dioxide in our atmosphere comes from volcanic eruptions.) Ultimately, plants can absorb the carbon atoms in carbon dioxide, at which point they are incorporated into glucose by photosynthesis. Sometimes animals that also are looking for food energy eat these plants. (Just think of the last time you saw a raccoon rummaging around in a garbage can, looking for last night's table scraps.) So a carbon atom that was once in a plant will now be in an animal, but still as part of a glucose molecule. The glucose may circulate around in the animal's bloodstream, to be taken up by one of the animal's trillions of cells to be used for energy. Or perhaps that animal's metabolism may pack it away in the liver as glycogen, a large storage granule composed of many thousands of glucose molecules for later use when energy is needed but the animal isn't eating.

This very process happens in us as well: if we eat three square meals a day, we eat more than we need during the moment of the meal or even for the next little while afterward, so most of the meal is stored as glycogen, a process regulated by insulin. Later on when we're not eating, that energy is accessed, and glucose is released from the larger glycogen stores. This allows us to avoid constant feeding so that we can do other things. (Obesity occurs when we tend to eat and store more than we'll need later and do so on a continual basis, with the excess continuing to store over time as fat.) Eventually, that glucose molecule will get used for energy, and the carbon atoms in it will be exhaled as carbon dioxide molecules through the lungs according to the respiration equation shown earlier, ultimately finding their way once more into the atmosphere, available to go through the cycle all over again.

Carbon atoms can also recycle through the planet in other ways, however. For example, all animals die eventually. When animals die, they have a lot of carbon inside them, the remnants of all of the chemical processes that were happening while they were living. If an animal carcass slowly changes via bacterial decay (instead of being eaten, in which case the carbon simply becomes part of another animal to carry on the cycle), that carbon will eventually become part of the soil, on a forest floor, for example. It could remain there for many years, but at some point, the soil will be disturbed and mixed up, perhaps by an animal such as a fox rooting around looking for his own food. The fresh soil that had been buried for years would then be exposed to the oxidizing effects from the oxygen in the atmosphere, and once again, carbon dioxide would be produced and released.

In special situations many millions of years ago, some plants would find themselves on a different path in the carbon cycle. If plants lived and died adjacent to swamps, in low-lying areas that were wet most of the time, their materials would accumulate. The effects of bacteria could cause the dead plant material to decay, sometimes becoming something known as peat. Peat requires conditions that are warm enough and wet enough to form, and ancient swamps from millions of years ago were just perfect for that. Peat is still used in some parts of the world as a source of fuel after it is first allowed to dry since it contains a lot of material that is good for combustion. But it can become an even better source of fuel for combustion given the right set of circumstances and enough time.

As material continued to accumulate millions of years ago and the depth of peat became greater, *anaerobic* conditions began to take place. The term "anaerobic" simply refers to a lack of oxygen; oxygen wouldn't be present deep under the ground's surface. Oxygen would normally help to degrade and decompose these materials if they were on the surface, but once anaerobic conditions were present, different chemical reactions took over. These particular reactions are determined by certain temperatures and pressures that are present when the conditions are just right, as they would have been in many places on Earth those many millions of years ago. These reactions led to a process known as coalification, and that term alone is likely enough to tell you where this story is going. Given enough time and the perfect set of circumstances, peat can be converted into something known as *lignite*. Lignite in turn can become *bituminous coal*—so called because of the tarry substance called bitumen that it contains, making it a softer form of coal—and that bituminous coal can in turn become *anthracite*.

These names may be unfamiliar to you, but they are the three grades of coal available, with anthracite being the highest quality because it's the most pure in carbon. The other grades have more impurities in them, containing other atoms besides carbon. Thus, ancient plant life, given the right conditions and the right length of time, ultimately became the coal deposits that provide much of the world's electricity today.

It's important to appreciate how long each of these cycles can take. Carbon dioxide in the atmosphere generally lingers there for hundreds of years. Once it's incorporated into the cycle of plant life with photosynthesis, it may remain there for many thousands of years. Carbon that is sequestered out of this cycle and into rock can stay there for many millions of years. But eventually, it will always find a way of moving from one source to another.

How Carbon Gets Around in Water

So far, the examples presented have referred only to plant and animal life on land, but most of the planet is covered in water. It shouldn't be too surprising to learn then that the same processes already described also occur in our oceans. Photosynthesis occurs in plant life, as you know, and the main plants found in our oceans are algae, which are very plentiful. (Indeed, most of the photosynthesis on the planet occurs in the plants in our oceans, with a much smaller amount produced by the plants on land. We just don't tend to think about plants in the ocean so much because we're land dwellers.) The carbon dioxide in the atmosphere can be absorbed into water itself, dissolving as a gas and then reacting with the water to produce carbonic acid.

$$CO_2 + H_2O \Leftrightarrow H_2CO_3$$

This is the same chemical equation that gives us carbonated beverages. We enjoy their slightly acidic taste, although much of the time they are artificially sweetened, such as in colas and ginger ale. (Beer is closer to the real taste of carbonic acid because it's unsweetened, and soda water is the best way to allow you the experience of what carbonic acid tastes like.) The unusual double arrow separating the two sides of the equation tells us that the chemical reaction easily goes back and forth between carbon dioxide and water on one side and carbonic acid on the other, driven by how much of each item in the equation is present. If carbon dioxide is added, and there is enough water around, then more carbonic acid will be produced, driving the equation to the right. If carbon dioxide escapes from the situation, then more carbonic acid will break down, and the equation will be driven to the left. In an unopened can of pop, for example, the reaction is balanced, but once a can of pop is opened, it will slowly lose its carbonation as the reaction starts to shift toward the left. It's also the reason we have to burp after drinking pop because the bubbles of carbon dioxide are being produced through this same process.

If the equation is being driven to the left, then the carbon dioxide can be released back into the atmosphere as a result, whether the can of pop sits on a counter and slowly goes flat or is consumed on a sunny afternoon on the back porch and the person drinking it eventually burps. (Shaking a can of pop will provide the carbonated water some additional energy, encouraging the equation to shift to the left even more, so that there's less carbonic acid and more carbon dioxide and water. That's why a shaken can will fizz so dramatically once it's opened.)

The dissolved carbon dioxide in the ocean is available for all aquatic plant life to use in the process of photosynthesis, producing carbohydrates such as cellulose and sugars, just like on land. And some of these plants are eaten by fish and marine mammals, just like they're eaten by animals on land. In turn, those predators use the carbohydrates for what they need, whether as food or to provide the material for the animal's structure, such as the protective shells of marine animals that are largely made of calcium carbonates. (This is similar to the outer shell of an egg, which is made from the same material.)

Just like on land, this carbon can also be released back into the ocean through the processes of respiration or animal death and decay. However, every so often, some aquatic animals die and fall to the ocean's floor without being eaten by other animals. This can be the beginning of a slightly different pathway for carbon than those already described. Marine life that falls to the bottom of the ocean after death can become part of the sediment on the ocean's floor, where it can remain for millions of years.

The ocean floor moves extremely slowly because it sits on something called a *tectonic plate*. It takes millions of years for even the smallest changes to occur because all geological processes take a long time compared to the time scales we use in our relatively short human existence. The surface of the Earth is covered by a series of these tectonic plates making up the Earth's crust; think of them as large masses of land floating on the hot liquid magma underneath. Oceans exist wherever water lies on top of these plates, but if you go down deep enough anywhere on the planet, you will eventually reach ground, and that will always be part of a tectonic plate. These plates have been moving around for billions of years at an extremely slow rate, but if you could watch a time-lapse movie of their movements, it might look something like bumper cars bouncing around haphazardly with no particular rhyme or reason. Some bump into each other, some slide against each other, and some pull away from each other.

Many of our planet's most interesting geological features result from the ways that tectonic plates can interact with each other. For example, where two plates have been colliding, land masses are pushed up against each other, building up to create mountain ranges. One of the most brilliant examples of this is where the Indian plate has been in a prolonged collision with the Eurasian plate, producing the Himalayas, which include Earth's highest peak, Mount Everest.

The Himalayas were created by the slow collision of two tectonic plates.

Where two plates rub against each other, with one moving in one direction and one moving in the opposite direction, a fault line is created. This leads to frequent earthquakes that result from the built-up tension and sudden release caused by the movement of these two plates grinding against each other. The San Andreas Fault between the North American and Pacific plates is the reason California experiences so many earthquakes. And where two plates are pulling apart, the hot magma beneath erupts as lava through the gap created between them. Once the magma cools, it hardens and becomes solid, essentially adding new material to the crust. An example of this is located in the middle of the Atlantic Ocean at the ocean's floor, where the eastern African plate and the western North American and South American plates are slowly pulling apart, creating the Mid-Atlantic Ridge. The African plate continues to move eastward, the North American and South American plates continue to move westward, and the ridge continues to build from the hot magma below.

As noted, on land tectonic plates can push against each other and build upward into mountain ranges, such as the Himalayas. In the ocean, however, an oceanic plate can collide with either a continental plate or another oceanic plate. In these situations, one plate can move under the other in a process known as *subduction*. One result of this is that material that was once on the ocean floor slowly becomes buried deep inside the Earth because the surface of one tectonic plate slides below the other. All of the sediment sitting on top, which is in part composed of the dead sea life that previously fell to the bottom, is carried along with the tectonic plate during this subduction, and it too becomes buried deep inside the Earth. This process occurs very slowly, taking many millions of years, just as all geological processes take a long time.

So the carbon from long-dead sea creatures stored on the ocean's floor has a method of becoming deposited deep inside the interior of our planet. It's an environment that provides plenty of heat because the centre of the Earth is a hot place. (Just consider what lava is like when it makes its way from the Earth's interior to the surface through a volcano.) There is also a tremendous amount of pressure because one tectonic plate on top of another means the weight of mountains is literally sitting on top of these sediments. With time, on a geological scale, the organic sediments undergo a number of chemical processes that convert them into something else. Depending on what the rock is like where this process occurs, how hot the temperature is, and how high the pressure gets, they may ultimately become liquid or gaseous.

Can you guess what these sediments might become? After millions of years, the liquid can turn into petroleum or crude oil, and the gas can become natural gas or methane, two major sources of energy that we drill for deep inside the Earth today. They are the same fossil fuels we use everywhere on the planet. It is precisely these geological processes that produced the materials that have been providing most of the energy our civilization has been using for the last few centuries. They are called fossil fuels because they are the organic remnants of plants and animals that lived on our planet, primarily in our oceans but also on land, many millions of years ago. They are a normal part of Earth's carbon cycle, and through the natural geological processes that occur, such as raging volcanic activity, some of those stored sources of carbon still make their way back into the atmosphere naturally. Nothing stays put on Earth forever.

Human activity has altered this carbon cycle, however. By using fossil fuels as the primary sources of energy, we have sped up the rate at which the carbon sources deep inside the planet make their way into the atmosphere naturally. We put carbon dioxide into the atmosphere at a rate many times faster than Mother Nature did before we came along. That is bound to lead to some changes.

The next chapter reviews our use of fossil fuels—coal, oil, and natural gas—all of which were created by these geological processes, transforming our planet from a simpler time before the Industrial Revolution into the technological wonder it is today.

Key Concepts of Chapter 2

- Given time on a geological scale, atoms and molecules move about the Earth easily from one region to another through a variety of natural processes.

- The movement of carbon as part of both organic and inorganic molecules occurs in a complex series of chemical and physical processes known as the carbon cycle.

- Fossil fuels were created through a variety of processes over millions of years to convert dead plant and animal material into coal, oil, and natural gas.

Suggested Reading for Chapter 2

Dressler, Andrew, and Edward A. Parson. *The Science and Politics of Global Climate Change.* New York, NY: Cambridge University Press, 2010.

Field, Christopher B., and Michael R. Raupach. *The Global Carbon Cycle: Integrating Humans, Climate, and the Natural World.* Washington DC: Island Press, 2004.

Wigley, T. M. L., and D. S. Schimel. *The Carbon Cycle.* New York, NY: Cambridge University Press, 2005.

Section 2
The Problem

Chapter 3
Harnessing Energy and How Fossil Fuels Are Used

> There is a sufficiency in the world
> for man's need
> but not for man's greed.
> —Mohandas K. Gandhi

Any form of energy we use needs to fulfill certain criteria to be practical on a global scale. If you were assigned the task of designing the most efficient form of energy imaginable to use on our planet, what are some of the characteristics you would have to keep in mind?

Cheap is a must—the global economy is driven by the possession and control of sources of energy. (This is certainly one of the factors we seem to value most in an energy source, as demonstrated by the fact that very few people are prepared to switch from something cheaper, such as fossil fuels, to something more expensive, as greener sources of energy tend to be, even if it's better for the planet in other respects.) You would want it to be energy-dense, which refers to getting a lot of energy out of a small unit of volume. Otherwise, it would be difficult to use in objects that are mobile and carry their engines with them, such as cars and planes. And you would want it to be easy to store and transport since energy use is needed everywhere, not just where it was obtained in the first place. Finally, you would want to have an endless supply of it, with no possibility that it will run out anytime soon. Ideally, you would also want it to be clean without any adverse effects on either people or the environment. Here's a summary of the factors to consider in an ideal energy source:

- Inexpensive
- Energy-dense
- Easily stored
- Easily transported
- Endless supply
- Clean

Fossil fuels such as gasoline and coal have many features on the list, in fact all but the last two. They are certainly cheap. The next time you are inclined to complain about the rising price of gasoline, think about it this way: the amount of energy you get from a gallon of gasoline is equivalent to what a human being can do with his or her bare hands in the form of manual labour in about 400 hours. (Think of how far that gallon of gasoline will take your family in a car, and now imagine pushing that car filled with your family the same distance by your own sheer brute strength!) At minimum wage, that gallon of gasoline is estimated to be worth more than $3,000 in man-hours. In that context, it's extremely cheap.

Fossil fuels are also energy-dense: the amount of energy you can get from a gallon of gasoline is significant for the volume it occupies. A battery would have to be much bigger in volume to contain the same amount of energy, at least with current battery technology. A wind turbine or a solar panel would have to be larger still. And fossil fuels are definitely easy to store and transport. Depending on where you live, you may not necessarily see too many oil pipelines, but trains carrying carloads of coal have likely held you at a railroad crossing at some point in your life, and if you live near a commercial seaway, you've probably seen cargo ships loaded with coal or oil go by.

Fossil fuels have the first four features listed that you would want in a source of energy, better than anything else we've come up with so far. But fossil fuels aren't renewable. Given the millions of years required to produce them, they will run out eventually based on our ever-increasing rate of consumption over the last few centuries. And of course, there are the adverse effects of using them, including air pollution, acid rain, respiratory health problems, and greenhouse gases (a topic discussed in much greater detail in the next chapter). To be fair, the human species decided long before you and I were born that fossil fuels would be the backbone of our global economy; also, the negative aspects of their use were not really understood when they were first being used, at least not to the extent that we understand these issues today.

Energy Sources in Days of Old

To understand how fossil fuels have transformed our world, first try to imagine living on Earth about three hundred years ago. Think of any pioneer village you might have visited, perhaps on a school trip. In that bygone era, you likely would have lived on a farm because most people were agrarian at that time. Families had to be self-sufficient, little microcosms of civilization; producing your own food was essential. To do this you would have lived in a farmhouse that you built yourself, or perhaps your father or grandfather built with the intention that it be passed down from generation to generation. The women in the family would have sewn the clothes you and your family wore. (My apologies to those offended by any gender inequalities in that statement, but it was a simple fact of the times that the tasks requiring less physical strength were usually handled by the women of the family; this included cooking, cleaning, and sewing.) The heat needed to keep your farmhouse warm and for cooking would have come from wood-burning. The light needed at night would have come either from candles made out of animal fat or beeswax or possibly from oil-burning lamps, with plant oils, such as olive oil, commonly used.

Manual labour was the way that things got done on the farm in those days, so you would have milked the cattle, slaughtered the pigs, and shorn the sheep all by hand. If your farm was near a river, you might have had a waterwheel to help generate some additional energy for you and your family. You even might have had a windmill if you were fortunate enough to have access to some reasonable winds. You would have tilled your fields with the help of beasts of burden, such as oxen. And if you wanted to travel to visit the nearest town, you would have used horses to get there, most likely riding horseback, although if you were a bit more aristocratic, you might have been able to go in style with a stagecoach.

The way you lived would have been much the same as it had been for those who came before you, literally for hundreds of years. Whether someone from the eighteenth century was a farmer eking out a living or the king of a vast empire, the energy available to the population at large came predominantly from people and animals, wood and plant oils, and to a lesser extent wind and water. Fossil fuels, so prevalent in today's society and on which our current social, political, and economic standards are based, were untapped resources back then. Coal was known, for sure, but no practical way of harnessing its power existed.

The Industrial Revolution

All of that changed with something known as the steam engine, invented nearly 250 years ago. The steam engine produces mechanical work such as an up-and-down or back-and-forth motion, with the energy from steam pushing on the engine's parts and making them move. Although there had been earlier versions, the one invented by James Watt around 1770 became the most successful, partly because it used the combustion of coal to produce the steam more efficiently than earlier models, and partly because it incorporated a modification that produced a rotary motion rather than simply back-and-forth or up-and-down. This was much better suited for use in factories.

The steam engine was able to extract the energy stored in fossil fuels on a wide scale for the first time in our civilization's history. The coal was burned in a combustion chamber and provided the heat to convert water into steam in a boiler. The expanded steam pushed against pistons, providing the energy to power the machinery and turn a wheel. The turning of a wheel produces a rotational energy similar to that created by wind turning a windmill or a moving river turning a waterwheel, but in this case the process was no longer dependent on a windy day or a raging river, both of which are far less reliable or consistent. Once the steam had cooled and used up most of its energy, it would no longer have enough pressure to be useful and would be vented into the atmosphere.

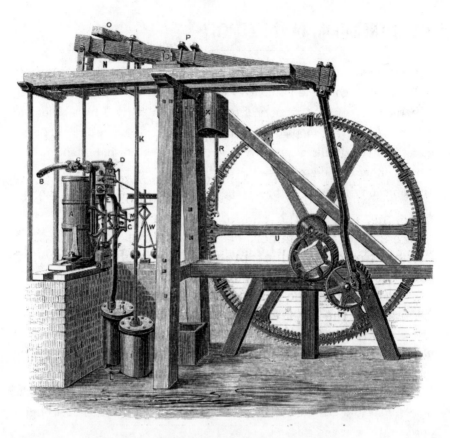

The steam engine that led to the Industrial Revolution.

Suddenly, thanks to this groundbreaking invention, a ready source of energy could be applied to various uses that made living much easier, including pumping stations, mills, and factories where textiles could be manufactured more easily than when looms were simply worked by hand. As the technology continued to evolve and improve, the principles of the steam engine led to the invention of the locomotive, transforming the world into a place where transportation across vast distances became a reality. The development of the steam engine truly heralded the onset of the Industrial Revolution and placed humanity on a path from which we have never looked back. It's a course from which we have never veered and from which we continue onward at a pace that makes it difficult to imagine anything coming along that will lead us in a different direction—other than perhaps a global crisis.

The Generation of Electricity

The steam engine just described may seem rather antiquated, but compared to livestock or good old-fashioned elbow grease, it was a much better way to harness energy. Today, most of the combustion of fossil fuels is done to generate *electricity* as opposed to generating mechanical energy directly. Before going into the details of precisely how that is done, it's important to have a basic understanding of just what electricity is. Most simply put, it is the phenomenon that exists in the presence or flow of an electric charge through a material such as a metal wire.

An electric charge occurs when there is an abundance of either positive atomic particles like our old friends the protons or negative atomic particles such as electrons. When these charged particles move through something like a wire, electricity exists. Some of these processes occur naturally, such as with lightning, static, or electric eels, but we also found a way to create the movement of these particles whenever we wanted to, thanks to the invention of the battery in 1800 by Alessandro Volta. Like many inventions, this creation followed the hard work of many scientists who came before Volta, including Benjamin Franklin and Luigi Galvani. (Notice anything interesting about the names Volta and Galvani and a connection to electricity? They are the origins of the words volt and galvanic.)

Batteries generate energy through a controlled chemical reaction. You will recall a description in chapter 2 of the chemistry of carbon and how carbon can be converted to carbon dioxide in the presence of oxygen via a chemical reaction. All of these reactions involve the movement of electrons from one source to another. If something is *oxidized*, it loses electrons. An example of oxidation is the corrosion of metal. When rust forms on cars, electrons are taken away from the iron by oxygen in the air. The iron is oxidized, producing a new chemical compound known as iron oxide, the proper chemical name of rust. The oxygen in the process gains the electrons lost by the iron and is *reduced*. The oxidized iron and the reduced oxygen together create an entirely new substance as a result of this particular chemical reaction. (A reduction-oxidation reaction is called a redox reaction for short.)

In a chemical reaction such as this one, in which iron and oxygen combine to produce rust, the movement of electrons is direct from one element or molecule to another. If, however, you could somehow control the reaction so that the electrons took a detour through a wire, that would create an electrical current. A battery does precisely that. It is nothing more than a controlled chemical reaction, with the flow of electrons generating electricity. In other words, batteries are a form of electrochemical energy. The first batteries involved placing metals in separate chemical solutions with wires connecting the two. Modern batteries still use the very same principle. The two metals in a NiCad battery are nickel and cadmium, for example. The chemical reaction will continue to run and produce energy as long as there is a wire connecting the two metals (such as when the switch on a flashlight is on), and as long as there are enough of the chemicals remaining to generate the electrons. Once they are used up and the redox reaction is complete, the battery has run out.

Electricity and Magnetism

The next step in our understanding of electricity deals with the special relationship between electricity and magnetism. You may have heard of the *electromagnetic* (or EM) spectrum before. In fact, all forms of visible light make up a particularly important part of the EM spectrum that we experience every day without thinking about it. Examples include the light you get from a light bulb or the sun. (The EM spectrum gets a lot more attention in the next chapter.)

There is no doubt that electricity and magnetism share an important relationship. But what is that relationship exactly? In 1821, Hans Christian Orsted discovered that as electrons travelled through a wire—what we have already defined as electricity—a magnetic field was generated around the wire as a direct consequence of the electrical current. It turns out the opposite is also true: if magnets revolve around a wire, they will induce electrons to move within the wire and, thereby, generate electricity.

This discovery led Michael Faraday to invent the electric motor soon after. His motor uses electricity as its power source, usually from a battery, creating a magnetic field when an electric current runs through a wire. Magnets placed at right angles to that current will be affected by it and as a result will be forced to move accordingly. The rotation that results from this motion can be used to move something, like any of the numerous battery-powered children's toys that we all grew up with. (My personal favourite was the slot car track!)

Since Orsted learned that electricity creates a magnetic field, and a magnetic field creates electricity, it makes sense that the electric motor that Faraday invented can be reversed as well. In other words, if some external power source rotates the magnets, then electricity will be created in the wire because the entire thing runs exactly opposite to the motor. So if we had something to turn our magnets and rotate them on an axis, then we would generate an electric current. To do this, we need something called a *turbine*.

A turbine is simply a shaft or drum with blades on it, similar to fan blades. These blades can be pushed by something moving past or through them, such as water or steam. Alternatively, pistons can be attached to the shaft, and steam can push on the pistons, which will rotate the turbine that way. This is only a small step beyond Watt's steam engine; in that case, steam was used to create a rotary motion, but that motion was used directly, such as in mills or locomotive engines. But with a turbine, the rotary motion is used to generate electricity. If there are magnets on the outside and wire coils on the inside of the shaft that's rotating, then lo and behold, electricity is produced.

Obviously, there are numerous power sources that can be used to get the turbine moving in the first place. Water (hydroelectric power, or hydro for short) and wind (as in wind turbines) are natural sources of energy. Nuclear power creates heat through nuclear reactions, and that heat can boil water into steam, subsequently turning a turbine. Most of the world's electricity, however, comes from the burning of fossil fuels. What it comes down to is simply that if you can't use something natural to produce a rotary motion, such as wind or water, then you have to do it by generating steam to turn a turbine for you, and that requires heat.

One drawback with these methods is that a lot of the heat that's generated is actually wasted rather than used to move the turbines. Nuclear power probably has an efficiency of 60 percent, and fossil fuels have efficiencies that are closer to 40 percent. The three main fossil fuels we use were described in chapter 2, namely, coal in solid form; petroleum or oil in liquid form; and natural gas, also known as methane, in gaseous form. Coal and oil are the two most commonly used fossil fuels today, accounting for more than a third each. Natural gas is in third place at about 25 percent of total usage, but it has been steadily climbing over the last few decades.

How we obtain these fossil fuels to use for energy is relatively straightforward: we mine for coal, and we drill for oil and natural gas. We've been mining for coal and drilling for oil in earnest since the eighteenth century, thanks in large part to the technological progress that resulted from Watt's steam engine. Initially, the natural gas that was often found in conjunction with oil wasn't considered of any value and was generally burned off at the source as a useless by-product. With time, however, those who were drilling learned that it too could be put to good use—with more profits to boot!

So how does the use of fossil fuels affect our old friend the carbon cycle? By burning them to produce the energy we need, we are altering the natural circulation of carbon, changing what Mother Nature has been doing for the last 500 million years. As you are already now aware, a great amount of carbon has been trapped inside Earth for a very long time, not just within the rocky crust but also as coal, oil, and natural gas. On a geological time scale, that carbon eventually would have made its way back into the atmosphere, either through tectonic-plate activity or volcanism, but over a few millions years. We've altered that, however, by removing this trapped carbon from inside Earth's crust and burning it as our major source of energy. In this way, carbon dioxide is getting into the atmosphere much faster than ever before, when it did so by way of the natural processes alone, as described in the last chapter. (Estimates suggest that the rate at which we are putting carbon dioxide into the atmosphere is at least twenty times the natural rate, that is to say the rate prior to the Industrial Revolution.) So when people say that by burning fossil fuels, we are simply doing what Mother Nature had been doing all along, that isn't so much in dispute. The critical difference, however, is the *rate* at which these changes are occurring. How quickly the planet can adapt to these changes is what truly defines the potential for global warming and a possible climate crisis.

How Fossil Fuels Differ

It's important to understand some of the differences among these fossil fuels. You may have heard that coal is the worst for the climate, and natural gas is the least offensive, at least as far as greenhouse gas emissions go. A little bit of basic chemistry explains why.

An important principle in chemistry is that different atoms are joined together by chemical bonds. It takes energy to break a chemical bond, but we get energy from a molecule when a chemical bond is formed. That's why so many complex molecules have formed all over the universe so easily. Atoms are naturally attracted to each other and readily bond together. So hydrogen is H_2, oxygen is O_2, carbon dioxide is CO_2, and water is H_2O. If we want to break a chemical bond, it takes some energy to do it. However, new molecules will be created because new bonds will be formed; the atoms that are now on the loose as a result of the breaking of those original bonds will want to combine with something. If the newly formed bonds produce more energy than it took to break the original bonds in the first place, then you'll have a potential source of energy.

Think of it this way: if you're at the cottage and you want to start a bonfire to roast marshmallows, you'll get a bunch of wood and some kindling to help the wood catch fire, and you'll start the whole process by striking a match. Energy first enters into the equation when the fire from the match starts the ball rolling. That lighted match is then used to burn the kindling, and this in turn ignites the wood. Wood is a good source of carbon, as all organic matter tends to be. (The term "organic" here refers to the fact that it came from things that were once alive.) So chemical bonds are broken by the energy released from the kindling and the wood, but new chemical bonds are formed as the carbon within these materials combines with oxygen from the air to become carbon dioxide.

This particular chemical reaction releases energy in the form of fire, providing both heat and light. The fire will break even more chemical bonds with its energy, but new bonds are again formed, releasing even more energy. As long as you have the fuel—in this case firewood—it will be a self-sustaining chemical reaction, with the energy from your bonfire ultimately coming from the process of breaking and reforming chemical bonds, releasing the energy that a carbon-containing fuel such as wood has to offer. (As an aside, note that atoms in molecules are held together by *chemical* bonds; energy is needed to break them, and they release energy when they are formed. Protons and neutrons within the nucleus of an atom are held together by *nuclear* bonds, and energy is needed to break them, and energy is released when they form as well. The energy from nuclear bonds is dramatically greater than that from chemical bonds and explains why an atomic bomb is so much more destructive than a conventional chemical bomb.)

Now let's turn this understanding of chemical bonds and the energy they can provide to fossil fuels and review each of them. First, we have coal: in chapter 2, we saw that there are different grades of coal, from the lowest grade, lignite, to the highest, anthracite. The higher the grade, the more pure it is in carbon, with less of the other types of atoms that can be found in coal, such as sulfur, hydrogen, oxygen, and nitrogen (all useful components of the chemicals of life, mind you). For the purposes of understanding, it is simplest to just think of coal as a big lump of carbon and nothing else. (The highest grade, anthracite, still has these other atoms in the mix, but about 95 percent of it is carbon.) If carbon in the form of coal is burned by combustion in the presence of oxygen, then carbon dioxide is formed.

Combustion of Coal: $C + O_2 \rightarrow CO_2$

The energy required to break up the bonds in the oxygen molecule is more than offset by the energy released with the creation of a molecule of carbon dioxide. As long as there is more coal to burn and oxygen is present, this is a simple way to obtain energy.

Carbon has a natural tendency to bond to other atoms, which is one reason it's so vital to the processes necessary for life. Carbon is able to form four bonds at once. (In the case of CO_2, it uses two bonds to each oxygen atom, so these are called double bonds. These are naturally much stronger than single bonds and are among the reasons that burning carbon and creating CO_2 is such a handy way to get energy—the energy is derived from the creation of four new bonds among only three atoms.) Carbon will frequently bond with hydrogen atoms because both elements have always been so plentiful throughout the universe, so the simplest of these types of molecules, known as *hydrocarbons*, is one carbon atom bonded to four hydrogen atoms. This is CH_4, also known as methane or natural gas.

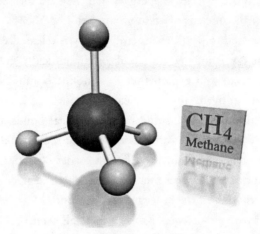

A molecule of methane.

If instead one of the bonds was to another carbon atom, but the other three were still bonded to hydrogen atoms—and if the other carbon atom did exactly the same thing—you would have a molecule of C_2H_6, or ethane. Adding carbon atoms to the chain would create even bigger hydrocarbons. Three carbon atoms would give you C_3H_8, or propane, commonly used in barbecues. Four carbon atoms would give you C_4H_{10}, or butane, the fuel used in cigarette lighters.

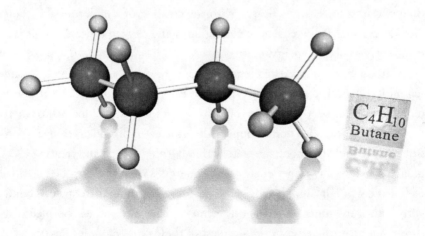

A molecule of butane.

And so it continues. Carbon chains can extend to great lengths. Eight carbon atoms give you octane, the predominant component in gasoline. Notice that lighter molecules with fewer carbon atoms, such as methane, occur in nature as a gas. Heavier molecules with greater chains such as butane and octane exist in liquid form. And, as you might imagine, even longer chains occur naturally in solid form. An example would be paraffin wax, which has a chain of twenty-five carbons. You also may have noticed that each carbon in the middle of the chain connects to two hydrogen atoms, and there are always two extra hydrogen atoms, with one at each end. These hydrocarbons follow a simple formula of $CH_2 + 2H$. The simplest is methane, CH_4. Octane is C_8H_{18}, and paraffin wax then is $C_{25}H_{52}$.

Methane, also known as natural gas, is frequently used to heat homes instead of electricity. Let's look at what happens when we burn methane as a form of energy instead of coal. In this case, we need to break the four bonds holding the carbon atom to four hydrogen atoms. This has to happen in the presence of oxygen, which is needed for combustion to occur, so the bonds connecting two oxygen atoms also have to be broken. There will now be some carbon, hydrogen, and oxygen atoms free to form new chemical bonds and create some new molecules, releasing energy in the process. Obviously, the carbon can combine with the oxygen to again make carbon dioxide, just as it did with coal.

But what does the hydrogen do? Since hydrogen and oxygen can easily combine together into H_2O, or water, that is precisely what happens. One molecule of methane and two molecules of oxygen can rearrange themselves into one molecule of carbon dioxide and two molecules of water. (All of the atoms on the left side of an equation have to be accounted for on the right side of the equation; remember that conservation of mass in chemical reactions is a law that cannot be broken.)

Combustion of Methane: $CH_4 + 2O_2 \rightarrow CO_2 + 2H_2O$

In the case of methane, it takes energy to break the bonds holding the carbon and hydrogen together, as well as the oxygen atoms in the oxygen molecules, but the energy produced in the formation of two water molecules and one carbon dioxide molecule more than offsets the energy that was invested in the first place. Just like our bonfire at the cottage, as long as you keep providing the fuel in the form of methane and initially start the whole process by investing a little energy with a spark or flame, then the energy output will be enough to provide a significant source of heat, perhaps to heat your home or cook your meals, if you use natural gas for those purposes.

An important factor to appreciate is that when we burn coal, all of the energy is generated from the formation of carbon dioxide. Burning natural gas, however, derives a lot of energy from the bonds made to form water as well. It isn't producing only carbon dioxide, as is the case with the combustion of coal. This is why coal is considered the dirtiest fossil fuel, since all it produces is carbon dioxide, which is then dumped into the atmosphere. Natural gas, on the other hand, is the cleanest of the fossil fuels because it produces two water molecules for every carbon dioxide molecule, and water can go almost anywhere on Earth to become ice, liquid water, or water vapour. Burning petroleum products such as octane also produces water along with carbon dioxide, but in ever-decreasing ratios. The longer the chain of carbon atoms in a hydrocarbon molecule, the closer the ratio of hydrogen atoms to carbon atoms will be 2:1, and the closer the combustion products ratio of H_2O to CO_2 will be 1:1.

That's why burning fossil fuels other than coal is considered cleaner, with natural gas considered the cleanest. Both oil and gas manage to provide some energy by creating water molecules compared with coal, which generates only carbon dioxide, and water isn't a big concern when it comes to harming the atmosphere. This is because carbon dioxide is a *greenhouse gas*, and water isn't, at least not in its liquid form. (Water vapour *is* a greenhouse gas, but the atmosphere can take only so much water vapour based on its temperature; any excess will become liquid water and precipitate out, so it's much less of a concern if we're creating water from these chemical reactions than if we're creating carbon dioxide, but there will be more on that in the next chapter.)

The use of fossil fuels has reshaped our world into one dramatically different from the one of our ancestors in their farmhouses three hundred years ago. With the Industrial Revolution, coal became a significant source of energy to help power steam engines. Petroleum products became a significant source around 1900, and natural gas about twenty-five years later. As industry continued to progress, with ever greater amounts of energy required to manufacture the things that our advancing civilization demanded, there was no turning back. Engines required for transportation in automobiles and planes took advantage of this energy source as well. And once plastics—which are produced from petroleum products— came on the scene, an even greater demand for fossil fuels occurred.

To this day, coal, oil, and natural gas are still the three most important energy sources on the planet. If you look at the trends over the decades, all three have consistently been on the rise, and it's easy to understand why, given an increasing global population and an exponential rise in industrialization. There have been some interesting blips in these trends over the years, such as the energy crisis in the 1970s, when the use of both petroleum and natural gas fell for a relatively brief period of time. But generally, their use has steadily increased and continues to do so with no clear end in sight. We use other forms of energy as well, such as nuclear and hydroelectric, but they are a very distant fourth and fifth place behind the big three.

This chapter began with descriptions of the features that would define a perfect source of energy. Fossil fuels have many of the characteristics we would like, but they have a few drawbacks as well. One referred to here is that they aren't renewable. The millions of years it takes to create them are much too long compared with the rate at which we're currently using them, so at some point in the future, they will run out. Much debate exists as to how soon that will occur, but it will happen someday. Another drawback with fossil fuels is the production of carbon dioxide, a greenhouse gas. The next chapter addresses how greenhouse gases impact our planet by affecting the atmosphere, ultimately changing the climate and contributing to global warming.

Key Concepts of Chapter 3

- Energy is needed to accommodate the demands of an evolving civilization.
- The energy sources that have been most ideal to date have been fossil fuels.
- Fossil fuels generate carbon dioxide in the combustion process, an important greenhouse gas.

Suggested Reading for Chapter 3

Downs, Jonathon. *The Industrial Revolution*. Westminster, MD: Shire, 2010.

Homer-Dixon, Thomas, ed. *Carbon Shift: How the Twin Crises of Oil Depletion and Climate Change Will Define the Future*. Toronto, ON: Random House Canada, 2009.

Kelley, Ingrid. *Energy in America: A Tour of Our Fossil Fuel Culture and Beyond*. Lebanon, NH: University of Vermont Press, 2008.

Remsen, Ira, and William Ridgely Orndorff. *An Introduction to the Study of the Compounds of Carbon or Organic Chemistry*. Charleston, SC: Bibliolife, 2009.

CHAPTER 4

Greenhouse Gases and How They Affect Our Planet

> It has become appallingly obvious
> that our technology
> has exceeded our humanity.
> —Albert Einstein

Think of the last time you got into your car on a hot summer day after it had been basking in the sun for a few hours. Remember how unbearably hot and stuffy it was inside, and how you immediately wanted to open up the doors and windows or get the air conditioning on, anything to cool the car's interior as quickly as you could. And if the car had a sunroof, and the window portion was closed but the visor had been left open, it was even worse. Have you ever thought about why that was? Why is it hotter inside a car than it is outside just because the car has been out in the sun? And why would it be a little cooler if the windows had all been left open? The answer has to do with something called the greenhouse effect. It's precisely this mechanism that allows a greenhouse used in growing plants to do what it does: keep the inside hotter than the outside. This principle also applies to what greenhouse gases do in Earth's atmosphere. To understand why greenhouse gases are potentially harmful to Earth's climate requires understanding this concept in a little detail.

Most people appreciate that the sun is the primary source of Earth's energy. It provides all of the light and almost all of the heat as well. (Earth has a little heat from its interior, which is still a hot place, as you are already aware.) What many people do not appreciate, however, is that the sun provides its energy to Earth primarily through light rather than heat. The sun is too far away—about 150 million kilometres or 93 million miles—for it to radiate heat directly through space and warm the Earth. That's just fine, though, because there's plenty of energy in light all by itself. Light energy is part of the electromagnetic (EM) spectrum, mentioned in chapter 3.

The Electromagnetic Spectrum

To understand the EM spectrum, it's first easier to think about sound waves as an analogy. Sound is made up of waves, and waves are simply vibrations with a crest and a trough, just like waves in the ocean. In the case of waves you might surf on, those waves are moving through water, but sound waves can move through all sorts of different substances. If you were to look at a tuning fork like one you might use to tune a guitar, you would probably see something printed on it that reads "A440." That means when you hit it against something, the tuning fork will vibrate at precisely 440 cycles per second (cps), giving a musical note. In this case the frequency of vibration produces an A note (thus the name A440). The tuning fork vibrates at 440 cps, which then makes the air around it vibrate at 440 cps, and that sound wave expands outward in the same way a pebble being dropped into a pond creates a ripple that expands outward in a wave. As the wave continues to expand, its energy gradually diminishes with the amplitude of the sound wave (the distance from the crest to the trough), gradually getting smaller and smaller. This is why sounds get softer the greater the distance you are from the source of that sound. But the wavelength (the distance from crest to crest, or trough to trough) doesn't change, so the pitch of the note remains the same. It is still an A note.

Sound waves expand just like ripples in a pond.

Once the wave of vibrating air gets to your eardrum, then your eardrum vibrates—again at 440 cps—which in turn makes the ossicles, which are the small hearing bones in your middle ear, vibrate. At what frequency? Well, 440 cps of course. The energy from that vibration leads to the vibration of fluid within your inner ear, causing electrical impulses in your brain to perceive the sound wave and appreciate the pitch of the tuning fork as an A note. If the tuning fork vibrated at a rate faster than 440 cps, the note would be higher. If the frequency were exactly double, it would be the same note but exactly an octave higher. Thus, 880 cps is also an A, but an octave higher than the A at 440 cps. If the fork vibrated at a slower frequency, then the note would be lower in pitch. If it vibrated at exactly half the frequency, 220 cps, it would be an A but an octave lower. The spectrum of sound waves is very wide, and there are frequencies above and below the normal spectrum of human hearing that still make sounds, just not those that we can easily hear. Think of a dog whistle that makes a pitch well beyond what humans can appreciate, but that canines can detect quite easily.

The EM spectrum is conceptually similar to the spectrum of sound waves. It covers a wide range of energies, and just as the sounds we hear are part of a wider spectrum of frequencies that exist, visible light is only a very small part of the EM spectrum, with many more EM frequencies both above and below those that our eyes can perceive.

Light is somewhat difficult to understand. In some ways it behaves like energy particles traveling through space. In fact, light behaves so much like particles that those particles have been given the name of *photons*. But light also behaves like waves, similar to the sound waves described previously. (Experiments have shown that light behaves like both particles and waves, and this is rather unimaginatively referred to as the wave-particle duality theory. For the purposes of this particular point, the wave theory is more appropriate to think about, but if you're interested, I would strongly recommend you pursue some further reading on this incredibly fascinating subject. We'll come back to photons and how they affect solar panels in chapter 6.)

Remember that with sound waves, different frequencies lead to different notes: the higher the frequency, the higher the note is in pitch. Light energy works similarly, but in this case a change in frequency affects the colour. A higher frequency will have a shorter wavelength, which you will remember is the distance from crest to crest (or trough to trough). The shorter the wavelength, the higher the energy level will be. In visible light, violet is at the higher end of the energy spectrum because it has a higher frequency and a shorter wavelength, whereas red is at the lower end of the energy spectrum because it has a lower frequency and a longer wavelength. (Light is usually referred to in wavelengths rather than in terms of frequency, but as you can easily appreciate, there is a direct relationship between the two. Light travels at a uniform velocity in a vacuum of about 300,000 kilometres per second, or 186,000 miles per second. The higher the frequency for a given colour of light, the shorter the wavelength needs to be to maintain that uniform velocity.)

We know that red is at one end of the visible light spectrum and violet is at the other. All remaining colours are intermediate between the two. From the high-energy end of the spectrum to low, we have violet, indigo, blue, green, yellow, orange, and red. Interestingly, white light is made up of the entire spectrum of visible light from one end to the other. A prism can easily refract white light, allowing the individual colours of the spectrum to be seen. A rainbow is one of Mother Nature's ways of doing the same thing, with the individual droplets of water collectively acting as one big prism.

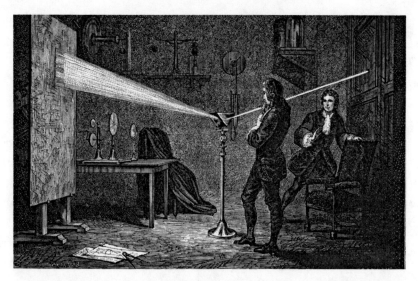

Sir Isaac Newton (1642–1727), English scientist and mathematician, using a prism to break white light into its spectrum. With Cambridge roommate John Wickins. Engraving of 1874.

The individual colours that we see when we look at objects depend on which parts of the spectrum are absorbed and which parts are reflected back. When white light from the sun shines on a blue car, we see blue because all other colours of the spectrum were absorbed by the pigment in the car's paint, and blue was the only one that managed to be reflected back to your eyes. When the chlorophyll in green plants is absorbing white light from sunshine to produce carbohydrates and oxygen through photosynthesis, it's absorbing all wavelengths of light except green which is being reflected back and perceived by your eyes as the colour green.

Why are our eyes so adapted to visible light rather than other parts of the EM spectrum? Because that's the most significant portion of the EM spectrum that manages to make it through Earth's atmosphere and reach the planet's surface. Much of the spectrum beyond visible light doesn't reach the ground because of the protective effects of our atmosphere. Components of our atmosphere prevent much of it from getting through and simply reflect it back into space. This is a good thing because much of the energy contained in the EM spectrum beyond violet is dangerous since the higher the frequency, the greater the amount of energy associated with it. Wavelengths with higher energy levels have the potential to disrupt the machinery located inside our cells and contribute to the development of cancer.

We do still experience some of those harmful effects because ultraviolet radiation is what causes our skin to tan or, if exposed for too long, to burn. *Ozone* is one of the protective components in our upper atmosphere, and it helps prevent too much ultraviolet radiation from getting to the Earth's surface, but not all of it. (Thankfully, chlorofluorocarbons that damage the ozone layer are less of a concern now than they were just a few decades ago, but more on that topic in chapter 7.) Putting it simply then, there isn't much point in being able to see anything outside of visible light if there isn't much of it to see down here on the ground anyway, so we've never evolved the ability to see beyond this narrow portion of the EM spectrum. But despite the fact that our best personal experience with EM radiation is with visible light, there is still a lot of the EM spectrum beyond it, and chances are that you've heard about many other components of this spectrum without ever realizing that they are all related, all part of the same form of energy.

First, let's look just beyond the spectrum of visible light. Obviously, we'll look figuratively rather than literally since our eyes can't perceive anything beyond the visible portion. Just beyond violet is ultraviolet radiation, referred to earlier. (Although most people call it ultraviolet *light*, it is more correct to refer to it as ultraviolet *radiation*, leaving the term "light" to that part of the EM spectrum that we can actually see with our eyes.) We can't see it, but we certainly notice its effects if we don't wear sunscreen. The energy contained within ultraviolet radiation is high enough to cause our skin to release melanin and become tanned or burned, depending on the duration of exposure. Even though many people view a tan as something healthy, tanning is our body's defense mechanism against the harmful effects of ultraviolet radiation, a higher-energy part of the EM spectrum than what our eyes can see. Just beyond red on the other end of the spectrum is infrared. *Ultra* means beyond, and *infra* means below, so the terms "ultraviolet" (just beyond violet) and "infrared" (just below red) make perfect sense.

Many people are already familiar with infrared radiation as well because it has to do with heat. As an example, heat-seeking missiles use infrared radiation to guide them to their targets. Modern barbecues often have infrared grills in addition to conventional ones. There's a simple experiment you can try at home that will demonstrate infrared radiation if you have a prism available, just like Sir Isaac Newton did in the late seventeenth century. When you shine a bright light through the prism, the spectrum of visible light will be displayed onto the wall with violet on one end, red on the other, and all other colours in between. If you then place a thermometer against the wall with the bulb immediately past the red portion of the spectrum that's displayed, the temperature will increase compared to the ambient temperature of the surroundings. Although there is no colour there, infrared radiation is still present because the prism refracts other EM energies, not just those of visible light. The thermometer will detect the heat from the infrared radiation.

The EM spectrum continues in both directions beyond visible light and even beyond ultraviolet and infrared radiation. In higher-energy wavelengths, there are X-rays and gamma rays. X-rays are used every day all over the world to image our bones and teeth, and with some impressive computing technology, they can image our internal organs to an amazing degree with computed tomography, also known as CT scans. Gamma rays are extremely high-energy wavelengths with very harmful effects to living tissue. That property is put to good use, however, as gamma radiation can be helpful in sterilizing equipment by killing harmful bacteria. It's also used in treating some cancers with a device known as a gamma knife, which is used to focus gamma rays on cancer cells, intending to destroy them in the process. At the other end of the spectrum, beyond infrared radiation, you'll find microwave radiation (which we use every day in our kitchens) and radio waves (which are used in radar, television, and obviously radio). Most of the EM spectrum is a part of our world, although visible light is the part with which we are most familiar.

The Greenhouse Effect

So what does all of this have to do with greenhouse gases? Let's return to our car's hot interior. Visible light from the sun reaches the inside of the car through the windows as white light. It hits the dashboard, the steering wheel, the seats, and anything else that happens to be inside. Whatever colour the interior is will be that part of the visible spectrum reflected back to your eyes while the rest of the light is absorbed. What happens to the energy from that absorbed visible light? It heats up the objects that absorbed it. An interesting property of any object is that it can radiate heat. You might not think about that very often because most things in our everyday world don't seem "hot." However, they all still have a temperature and will radiate that in the form of infrared radiation, increasing the temperature of whatever might be next to it. The hotter an object is, the more heat it can radiate. Although you can't see it, you can certainly feel it because infrared radiation will provide heat to anything it comes in contact with.

Notice that this heat is radiating back still within the EM spectrum, just no longer in the visible portion. Most of the light's energy was absorbed by everything inside the car, with only a little reflected back as colours. The absorbed energy was converted into heat that warmed up everything inside; the heat from the interior then radiated that heat to the air inside the car. And here's the nasty part of the whole story: the glass windows that let visible light in and started the whole process aren't so good at letting the infrared radiation out. Glass tends to reflect infrared radiation, so it bounces back, staying inside the car and continuing to contribute more heat. That's why leaving the windows open even just a little bit helps somewhat; the hot air inside has a chance to escape. It's also why the car is even hotter inside if the sunroof visor is open but the window is closed: doing that simply allows more sunlight to get into the car, subsequently converting even more visible light into infrared radiation that remains trapped inside because the windows aren't open to let the hot air escape so easily.

This principle makes for some unpleasant experiences in the car on hot summer days, but it's been put to good use with greenhouses. When visible light is allowed through the windows, which make up the bulk of a greenhouse, the plants inside can absorb the light energy, reflecting back their individual colours—most typically green, although other colours will be there too—but absorbing the rest and using some of that energy for photosynthesis. However, not all the energy that is absorbed is used for photosynthesis—plants are nowhere near that efficient—so in a manner somewhat similar to the interior of the car, a lot of the energy absorbed by plants is converted into heat, which is then radiated back in the infrared portion of the EM spectrum. And just as it was with the car's windows, the glass making up the structure of the greenhouse won't let the infrared radiation out so easily, so it gets reflected back. This keeps the greenhouse interior warmer than the outside, allowing plants that require a warmer environment than the outside temperature would normally allow to grow. In this way, we can grow tropical plants in more moderate temperature zones by using greenhouses, and plants can also grow all winter long.

As you can now likely anticipate, a greenhouse gas does the same thing as a greenhouse, or it wouldn't be called a greenhouse gas. In this case, greenhouse gases in the atmosphere are like the glass windows of a greenhouse: they allow visible light through, some of which is reflected back (white from clouds and the ice at the poles, blue from oceans, green from regions with plenty of plant growth, and yellow from deserts), but a lot of which is absorbed. Much of that absorbed energy gets converted into heat and radiated back as infrared radiation. And just as glass allows visible light through but reflects most of the infrared radiation back, greenhouse gases also let the visible light from the sun through but reflect much of the infrared radiation back, keeping the planet's interior warmer than it would be if not for this insulating blanket we have from our atmosphere. This is important for our planet's history because if it weren't for this process, it would be far too cold for us to live on Earth as comfortably as we do. We are adapted to live in a climate that depends on a certain temperature range, and that range depends on some of the heat that we receive from the greenhouse gases that are present naturally in our atmosphere.

The Composition of Earth's Atmosphere

This brings us to a very important point: greenhouse gases have been around much longer than we have. They aren't exclusively a product of human activity. But not every gas in Earth's atmosphere is a greenhouse gas. In fact, most of them aren't. Let's quickly review the gases that make up our atmosphere:

Nitrogen	78%
Oxygen	21%
Water vapour	1%
Argon	0.9%
Carbon dioxide	0.04%
Methane	0.0002%

Thus, a quick review shows that the bulk of our atmosphere—about 99 percent—is made up of nitrogen and oxygen. Nitrogen is relatively inert, and of course, oxygen is necessary both for animal respiration and for combustion, both of which have been described previously. But these two gases do not contribute to a greenhouse effect. In other words, only a very tiny fraction of the entire atmosphere consists of gases that contribute any significant amount to Earth's global temperature.

As for the rest of the gases found in Earth's atmosphere, about 1 percent is water vapour. Remember that water exists on our planet in three forms: water vapour, liquid water, and ice. Water is truly one of the defining and unique qualities of Earth, at least compared to any other planet in our solar system. Water vapour is water in a gaseous form, found in the atmosphere. You can easily see evidence of water vapour because it frequently condenses into clouds, and it's also easy to appreciate it on a muggy day; that heavy feeling in the air is from water vapour. The hotter the air, the more water vapour it can hold. That's why skies are so much clearer and the horizon so much more easily and clearly seen during winter months when the moisture content in the air is lower (a cruel irony to amateur astronomers, who get their best and clearest "seeing" during the coldest months of the year). For any given temperature, however, the atmosphere can hold only so much water vapour. If more water vapour is added, it will reach a saturation point and condense into liquid water, ultimately precipitating out as rain, or snow and hail if the temperature where it precipitates is cold enough.

Water vapour is an important greenhouse gas. Truth be told, it's the most potent because it has more of an effect on the greenhouse process than any other gas in the atmosphere, molecule for molecule. But if average temperatures worldwide are stable, water vapour is generally a constant because it is difficult to add more for the reason described in the last paragraph. (Of course, this raises an important question: would the atmosphere hold more water vapour if global temperatures increased? The answer is yes, and I will expand more on that in the next chapter.) Also, water vapour molecules generally linger in the atmosphere for only about ten days or so. This is quite brief compared to the other greenhouse gases, which are able to remain in the atmosphere for centuries. That's why human activity doesn't play a very significant part in water vapour concentrations directly: water vapour molecules added by us from the combustion of hydrocarbons don't tend to stay in the atmosphere for very long.

Argon is an inert gas that doesn't easily interact with anything. It's a member of a family of elements on the right side of the periodic table known as "noble gases," considered noble simply because they don't tend to react much with the other elements. There are six noble gases in total: helium (used in birthday balloons), neon (used in brightly lit signs outside many shops and restaurants), argon, krypton (which sadly has nothing to do with Superman—other than perhaps providing his creators Jerry Siegel and Joe Schuster with a cool sounding name for his home planet), xenon, and radon.

The next component on the list is carbon dioxide, making up just under 0.04 percent of the atmosphere. This is sometimes referred to as 400 parts per million (ppm) because it's a little easier to say—and to visualize for some reason—than 0.04 percent. In 2010, the measured amount of carbon dioxide in the atmosphere was about 388 ppm according to the Scripps CO_2 program, which has been making precise measurements of carbon dioxide for many years. Methane makes up an even smaller part of the atmosphere's composition, at a level of about 1,745 parts per billion (ppb). Methane is about twenty times more effective as a greenhouse gas than carbon dioxide (again molecule for molecule), but because the concentration of carbon dioxide is about 220 times that of methane in the atmosphere, methane makes a much smaller impact overall. There are other greenhouse gases in the atmosphere as well, such as nitrous oxide, ozone, and chlorofluorocarbons, but these are all much smaller players in the big picture and need no further discussion for our purposes.

The individual components that make up our atmosphere come from a variety of sources. The concerns about greenhouse gases arise because we create so much of them through our human activities. Combustion of fossil fuels is an obvious source, but there are some other ways we add to greenhouse gases that you may not have realized.

OTHER SOURCES OF GREENHOUSE GASES

There are methods by which we add carbon dioxide to the atmosphere besides the combustion of the three types of fossil fuels we use as our sources of energy. These consist of things we do to help us live our lives: namely, deforestation and agriculture. When space needs to be created for an expanding population to live in, it's often the case that significant forests are cleared out of the way. Just think about the city you live in right now. Has it experienced any growth since you first lived there? Have you noticed the removal of any trees to make room for new housing and commercial developments? I bet that you have. If you could go back in time a few centuries to an era before your city was developed to its present level, it's likely that you'd see many more trees located there. Most of the trees that were removed to allow your city to expand its borders simply would have been chopped down, removing important plant sources that help to remove carbon dioxide through photosynthesis.

Elsewhere on the planet, however, many forests are simply burned down to make way for growth. (Of course, forest fires occur naturally and have since long before the Industrial Revolution, so greenhouse gases can be added to the atmosphere through this mechanism naturally without our help. Unfortunately, higher global temperatures are making forest fires more common and more devastating than ever, so we've been aggravating the contribution from this source indirectly even further.)

In chapter 3, we reviewed the various fossil fuels that we burn to provide us with energy. But one of the earliest sources of energy we ever used was wood, and wood is another source of carbon, just as all organic matter is. Wood isn't as useful a source of energy as the fossil fuels are because it isn't as densely packed with energy; a certain volume of wood will not provide anywhere near the energy that could be obtained from the same volume of coal, oil, or natural gas. But because wood is a source of carbon nonetheless, deforestation by burning is a significant source of carbon dioxide that is added to the atmosphere. Some countries are burning down forests at particularly alarming rates. If you look at a picture of the border between Haiti and the Dominican Republic, it provides for an amazing contrast. Haiti has a different set of regulations about deforestation than the Dominican Republic, and it shows. In 1923, 60 percent of the land in Haiti was covered in forest, but by 2006 it was down to only a rather unsettling 2 percent.

An example of deforestation.

Agriculture is also an important source of carbon dioxide production. If you want to develop large acres of fields into farmland, usually you have to prepare the soil. (It's likely that some trees have to be removed along the way as well.) A simple process such as tilling the soil allows bacteria that were previously buried and dormant to be overturned. Bacterial activity can generate its own source of carbon dioxide by degrading and oxidizing the newly exposed carbohydrates with which the bacteria come in contact.

However, we also use livestock on this planet for food and clothing, so we have large herds of cattle and sheep. These animals are called ruminants, and they have four stomachs to aid in their digestion. Ruminants digest grass very slowly because cellulose is extremely difficult to digest. In fact, they have bacteria in their stomachs that allow them to do that because they can't do it on their own. A by-product of the digestion of cellulose is the production of methane, one of our greenhouse gases. Ruminants such as cattle and sheep belch the methane they produce, allowing it to be released into the atmosphere. Since there are many people all over the world who eat meat and consume dairy products that come from livestock, that makes for a whole lot of methane being belched out. There are about 1.3 billion cows in the world, about one for every five people on the planet. The US Environmental Protection Agency estimates that these cattle vent about 300 trillion litres of methane annually. This equals about 20 percent of all human-related methane emissions and rivals that produced by the natural gas and petroleum industries. In the last twenty years alone, methane levels have increased from about 1,650 to 1,775 parts per billion, in part because of the large number of livestock on our planet.

Thus, we have expanding populations that need places to live, so we remove forests to make space; we till soil to grow ever-increasing amounts of plants to eat; and we herd large numbers of cattle and sheep for sources of food, dairy, and clothing. And don't forget that once forests are cleared, buildings often go up that take energy to construct; once they're built, they take even more energy to light and heat. We also ship produce and animal meat all over the world, burning fossil fuels in the transportation process.

An expanding global population leads to an increase in greenhouse gases at every level. All of these various activities lead to increased emissions of carbon dioxide along with the other greenhouse gases. This raises the point that one way to help tackle the problem of global warming is to somehow curb the rise in the global population—easier said than done, but progress in the developing nations can help. This can include better sanitation, immunization, and health care. With an improved standard of living and lower mortality rates, parents won't need to try to have bigger families, something they often do because they generally anticipate a few deaths among their offspring. Also, improved education, equal rights for women, and the growth of democracy all have the potential to alter the continuing trend of a rise in global population.

To recap, the main greenhouse gases are water vapour, carbon dioxide, and methane. Water vapour has the biggest impact, making up a greater percentage of the atmosphere than all others combined, but its amount is generally constant for given temperatures, and whatever we add doesn't last long, usually only a matter of a few days. The other two, carbon dioxide and methane, are much smaller percentages, but they have some distinct differences from water vapour. They aren't constant because human activity leads to an increase in their amounts from the combustion of fossil fuels, deforestation, and agriculture. They also linger in the atmosphere much longer than water vapour molecules, for centuries rather than simply days.

The combustion of fossil fuels in effect reverses what photosynthesis has been doing for millions of years because photosynthesis stored energy in materials that ultimately became those fossil fuels. We've been burning fossil fuels in earnest ever since the Industrial Revolution, so you would expect that we have been adding carbon dioxide to the atmosphere for just a few centuries. And because carbon dioxide lasts in the atmosphere for centuries rather than just days, we might expect that the levels would be gradually increasing unless our planet has some natural methods to compensate for this. (One potential example of such compensation might be if photosynthesis was stepping up its rate beyond what it's been doing for millions of years in response to the higher levels of carbon dioxide and keeping them at a constant.)

How can we know whether carbon dioxide levels are changing? Simple: we can measure carbon dioxide's concentration in the atmosphere over time. As mentioned previously, measurements in 2010 placed the concentration at 388 ppm. Has it always been that concentration, or do we have any evidence that it's been changing?

Fortunately, we have an answer to that question. In the late 1950s, a scientist by the name of Dr. Roger Revelle wondered this very same thing, He hired a research partner, Dr. Charles David Keeling, when he worked at the Scripps Institution of Oceanography, and in 1958, they decided to start measuring carbon dioxide levels. Their plan was to follow the levels with time to see what they were doing. They chose to measure carbon dioxide concentrations at two spots on Earth: at Mauna Loa, an extinct volcano in Hawaii, and in Antarctica. They chose these particular locations because they knew that there would be little impact from local effects of civilization, so the results would be more reliable than if measurements were made in the middle of a big city. Mauna Loa is a high peak at almost 4,300 metres, or 14,000 feet, above sea level. Also, Hawaii is a relatively small land mass with little by way of major industry, so measurements taken at its highest peak should be relatively accurate. Antarctica would have no local effects from human activity for obvious reasons as well.

Dr. Revelle and Dr. Keeling's findings were eye-opening. Although they have both passed away, measurements have continued on with their successors through the Scripps CO_2 Program. If you look at a graph of their findings, you can easily see that there has been a steady increase ever since the measurements began, at a rate of about two parts per million per year on average. It's also easy to appreciate that there is an annual cycle. The peak of carbon dioxide concentration throughout the year occurs in the spring, and then the concentration begins a steady decline until the autumn, when it reaches its lowest point throughout the year. It then begins to climb again, reaching its peak the following spring to begin the cycle all over again.

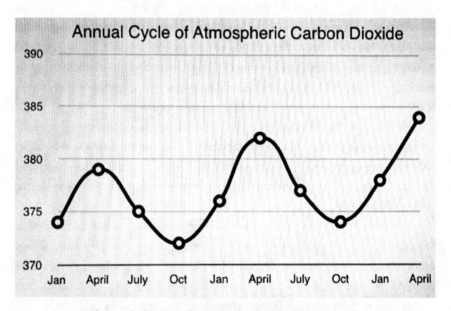

Carbon dioxide levels fluctuate naturally throughout the year.
(Graph by author, based on Keeling data)

Understanding this cycle is really quite easy. Recall that photosynthesis takes carbon dioxide out of the atmosphere to make carbohydrates and oxygen. When would that process get going to any substantial degree? Naturally, that would occur once the winter months are over and vegetation begins to grow. So you would expect that carbon dioxide levels would decline during spring and summer, reach their lowest point in the fall, and then start a steady rise until the following spring. That is exactly what occurs.

A smart question you might be asking yourself is, Why would that be, since when it's winter in the Northern Hemisphere, it's summer in the Southern Hemisphere? Wouldn't it all balance out? The answer is yes, it would balance out *if* the amount of land on Earth was equally divided between the Northern and Southern Hemispheres, but that's not the case. If you look at a map of our planet or a globe, you'll notice that most of the land masses are in the Northern Hemisphere. For this reason, the Northern Hemisphere's influence on carbon dioxide levels wins out, although there is more balance than there would be if absolutely no land existed south of the equator.

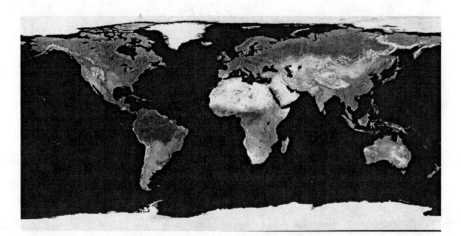

A map of Earth showing that most of the land is located in the Northern Hemisphere.

For the last half century, during which these careful measurements have been taken, carbon dioxide levels have been steadily climbing, something now described as the Keeling Curve. They go through their annual cycle of rising during the Northern Hemisphere's winter months and dropping during the Northern Hemisphere's summer months, but each year, the levels climb a little higher than the year before. Given that we've been burning fossil fuels steadily and in ever-increasing amounts ever since the Industrial Revolution began, producing carbon dioxide as a by-product, it makes sense that atmospheric carbon dioxide levels would increase.

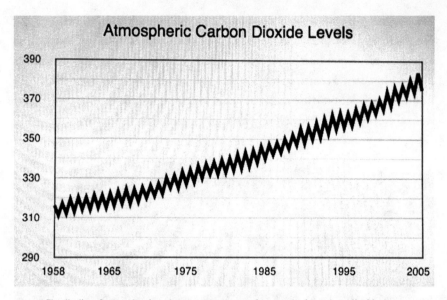

The Keeling Curve, showing that measurements of atmospheric carbon dioxide have experienced a steady increase over the years. (Graph by author, based on Keeling data)

But couldn't there be other forces playing a part in this? Might this trend be present even if human activity weren't contributing at all? We know what has been happening for the last half century since Drs. Revelle and Keeling started their measurements, when the amount of carbon dioxide in the atmosphere was about 315 ppm (Keeling et al., 538–551). But can we look back even further in time? Are there ways we can determine how long this trend has been going on?

It turns out that there are methods to confirm what levels of carbon dioxide, along with other greenhouse gases, have been in years past. If you wanted to know what the concentration of various atmospheric gases was a long time ago, is there any place on Earth you could look where those concentrations would be preserved as they were in the past? In other words, are there captured pockets of air anywhere we can study?

Indeed there are. When snow falls and accumulates, slowly compacting into denser snow and ice, little pockets of air get trapped along with the snow and ice. But of course, those pockets are released once the snow and ice melt—except where the temperature never warms up enough to melt them. In that case, a slow and steady accumulation continues, with trapped pockets of air from the time when the ice was formed. The deeper the level of ice, the further back in time you go. This is exactly what happens in the Antarctic, where the accumulation of ice has gone unhindered for thousands of years. By digging out a core of ice, scientists can analyze samples of the atmosphere that have been preserved to determine the concentrations of all of the components of the atmosphere at any time in Earth's past over thousands of years.

So what have we seen with the concentration of greenhouse gases that have been obtained from Antarctic core ice samples? Has a trend been continuing on for much longer than just a few hundred years? Is the climb part of some other process besides human activity? From the studies performed on these ice cores, it has been clearly demonstrated that three greenhouse gases, namely carbon dioxide, methane, and nitrous oxide, were at very stable levels for many centuries, but all three started a steady climb around two hundred years ago.

This makes for a strong case that human activities, such as burning fossil fuels, deforestation, and widespread farming and agriculture practices, are the likely culprit. As an ever-increasing global population requires places to live, food to eat, and an increased demand for energy on a per capita basis, we have added enough greenhouse gases to the atmosphere to explain the bulk of the trend that we've been observing; we can determine this directly for the last fifty years by measuring the actual concentrations in the atmosphere and indirectly in centuries past by measuring the concentrations in trapped air pockets within ice cores from the Antarctic.

Since we know that greenhouse gases get their name because they have an insulating effect on the planet, allowing the sun's light through but trapping the subsequent infrared radiation as heat, a steady climb in greenhouse gases would be expected to lead to a steady climb in global temperature as well. Have we seen that happening? Does proof exist that these emissions are actually changing our atmosphere, contributing to global warming? Chapter 5 addresses some of the evidence for a rise in global temperature and some of the effects we might predict as a result of that warming, along with changes we have already observed.

Key Concepts of Chapter 4

- A greenhouse gas allows visible light through but traps infrared radiation, insulating our planet as a result.
- Water vapour, carbon dioxide, and methane are three important greenhouse gases found in our atmosphere.
- Evidence indicates that carbon dioxide and methane had been at steady levels for centuries but have been increasing consistently ever since the Industrial Revolution.

Suggested Reading for Chapter 4

Association of American Geographers GCLP Research. *Global Change and Local Places: Estimating, Understanding, and Reducing Greenhouse Gases*. New York, NY: Cambridge University Press, 2010.

Flannery, Tim. *The Weather Makers: How We Are Changing the Climate and What It Means for Life on Earth*. Toronto, ON: HarperCollins, 2005.

Keeling, C. D., J. A. Adams Jr., C. A. Ekdahl, and P. R. Guenther. "Atmospheric Carbon Dioxide Variations at Mauna Loa Observatory, Hawaii." *Tellus* 28 (1976): 538–551.

Lovelock, James. *The Revenge of Gaia*. New York, NY: Penguin Books, 2006.

CHAPTER 5

Global Warming and Its Devastating Effects

> Only when the last tree has been cut down,
> Only when the last river has been poisoned,
> Only when the last fish has been caught,
> Only then will you find that money cannot be eaten.
> —Cree prophecy

How do you keep yourself warm when you retire to bed for the evening? If you're going to sleep on a chilly winter night (and you're doing your part to reduce your carbon footprint by turning the thermostat down), you can simply grab an extra blanket in order to be a little toastier. The more blankets you add, the warmer you'll be. Adding the extra insulation helps to trap the heat around you rather than letting it escape into the room. That's precisely what the addition of greenhouse gases does to Earth's atmosphere, effectively adding more blankets of insulation and keeping the global temperature warmer. In the last chapter, we reviewed some key evidence that carbon dioxide and other greenhouse gas levels have indeed been climbing as a direct result of human activities, such as burning fossil fuels, deforestation, and agriculture. Now let's look at some of the evidence showing that Earth's global temperature has indeed climbed as a result of those greenhouse gas emissions.

Reliable and accurate temperatures have truly only been available since the latter part of the nineteenth century. If we look to the last 150 years of accurately recorded measurements, there has been a consistent trend of increasing global temperatures. In fact, some of the hottest years on record have been just within the last decade. Since we have been steadily adding insulation to the planet through the production of greenhouse gases, the average global temperature on Earth indeed would be predicted to rise as a result of the insulating effects from those gases, so the observations make sense without needing alternate explanations. However, just as we looked to evidence of an increase in measured carbon dioxide that has been recorded over the last fifty years and found it in the Keeling Curve, it's not enough to simply notice a recent increase in global temperatures and assume that it results directly from greenhouse gases. It's necessary to prove it.

A Look Back at Earth's Temperatures

To learn about atmospheric concentrations of greenhouse gases, we had ice cores from the Antarctic to help. That research confirmed that the increase really began around the onset of the Industrial Revolution. Carefully detailed thermometer measurements are great, but how do we observe temperatures from years gone by? Are there any methods to determine Earth's temperatures from Two hundred years ago? Or even five hundred or one thousand years ago?

We obviously don't have something so precise as the preservation of temperatures like we have for trapped atmospheric gases from ice cores, but we do have a variety of methods where indirect evidence can be found that reflects our planet's past temperatures. The methods used in confirming Earth's temperatures before accurate thermometer recordings were available—in a branch of science known as paleoclimatology—are more complex than the intended scope of this book, but in general, warmer and colder periods in Earth's history leave a variety of telltale clues.

Here, we'll review one such clue as an example, again found in ice core samples. Just as these core samples trapped little pockets of air from many years ago, they also trapped water molecules from the same time periods. The different concentrations of oxygen *isotopes* found in these trapped water molecules reflect changes in global temperature at the time the molecules were trapped in the ice.

It's time to review what isotopes are exactly. Remember that all atoms have their atomic number determined by the number of protons contained in their nucleus, described back in chapter 1. The number of neutrons in the nucleus can vary, however, and thus so can the atomic weight of the atom. These differences in atoms of the same element are called isotopes. One example would be the two isotopes of hydrogen, with one having a neutron in addition to its solitary proton; when combined with oxygen, this version of hydrogen—known as deuterium—produces heavy water. Another example is the set of carbon isotopes, carbon-12, carbon-13, and carbon-14; all three forms of carbon have six protons, but one has six neutrons, one has seven, and one has eight. About 99 percent of all the world's carbon is carbon-12, with almost 1 percent carbon-13 and only very small trace amounts of carbon-14 found in nature. The first two isotopes of carbon are very stable, but the third is radioactive.

It's the property of radioactive decay that allows for carbon dating: after about 5,730 years, half of the carbon-14 will decay to nitrogen-14, with one neutron converting into a proton in the process. In other words, carbon-14 with six protons and eight neutrons becomes nitrogen-14 with seven protons and seven neutrons when it experiences this type of radioactive decay. After another 5,730 years, another half will have decayed even further, so that only about 25 percent of the original amount of carbon-14 will exist. This continues on indefinitely, leading to another reduction by half of the remaining carbon-14 every 5,730 years. Living organisms have amounts of carbon in these fixed ratios, maintained throughout life by the carbon that's ingested in food that also contains carbon in these same ratios. (Carbon-14 on Earth is replenished at a steady rate; cosmic rays from space create it when they hit atoms of nitrogen in the upper atmosphere, converting it to carbon-14 in the process. Since carbon atoms travel all over our planet, they gradually mix with all of the other carbon atoms found on Earth in the observed ratios described previously.)

Once an organism dies, however, the ratio is no longer preserved. Radioactive decay leads to a gradual decline in the amount of carbon-14 according to the precise rate just described, with a loss of 50 percent every 5,730 years. This period of time is known as its half-life. This fact allows us to determine the age of organic material that lived once upon a time. If the percentage of carbon-14 found in some discovered organic matter is only 25 percent of the usual percentage found in nature, then it's clear that two half-lives have passed since that plant or animal died, making it about 11,460 years old. Carbon-14 dating is useful for plants and animals that died up to about 60,000 years ago, or around ten half-lives ago. After that period of time, the amount of carbon-14 remaining is so minimal that accurate estimates are no longer possible.

Just as there are different isotopes for hydrogen and carbon, there are different isotopes of oxygen as well: oxygen-16 and oxygen-18. Since oxygen-18 has a higher atomic weight due to the two extra neutrons it contains, it takes more energy to evaporate water molecules containing oxygen-18. The higher the global temperature of the planet, the higher the average surface temperature of the oceans will be. If this is the case, then you might expect to see more water molecules containing oxygen-18 evaporating than when the temperatures are cooler. And that's precisely what happens. Higher global temperatures lead to a higher ratio of oxygen-18 to oxygen-16 in the atmosphere, and subsequently, that ratio is reflected in the water that's precipitated out and trapped in the ice that is layering in the Antarctic, available to be drilled as ice core samples many years later.

The changing ratio of oxygen isotopes is a reasonable sign to tell us what the temperature of our planet was long before accurate thermometer measurements were being taken. There are other clues in different regions of the planet that indicate what previous global temperatures have been as well. These include differences in ocean sediments and the appearance of tree rings on land and coral rings in the ocean. There are limitations with this kind of indirect evidence of course, but with the number of different methods available to paleoclimatologists, the agreement observed among these various methods is considered quite reliable.

Looking at this historical evidence, it's easy to note that global temperatures have had some fluctuations, primarily as a result of changes in our planet's orbit (explained in greater detail later in this chapter), but the overall trends have generally been quite constant over the last millennium, fluctuating within about a half-degree Celsius—constant, that is, until about the last 150 years, during which time the average global temperature has steadily climbed by almost 1 full degree Celsius. This is again rather convincing evidence that our human activities have led to an increase in greenhouse gas emissions, and they have been the main culprit in the observed increase in recorded temperatures since 1880, when accurate and reliable thermometer recordings began. We don't have to look to some as-yet-unexplained phenomena that are causing this increase in temperature. Like it or not, it's simply due to us and our need for growth: growth in population, growth in the area of land we live on, growth in agriculture and farming to feed and clothe everyone, and growth in the energy use to provide us with the quality of living we've become accustomed to and expect. This is what the science would predict, and it's exactly what the observations demonstrate.

With global warming, what might you expect to see? Some of the changes are rather intuitive, but others are not. Some of the more significant and potentially devastating effects include melting ice, increasing ocean acidity, severe weather changes, and the spread of disease.

Global warming will have devastating consequences for the environment.

Melting Ice, Rising Sea Levels, and Ocean Acidity

Warmer global temperatures will obviously lead to the melting of more ice, which will remain in the form of liquid water. Thus, polar ice caps will shrink, and glaciers will disappear. Do we have evidence that this has occurred? If you've watched the movie or read the book *An Inconvenient Truth* by Al Gore, then you've seen a number of dramatic examples of these sorts of changes. Indeed, they're quite startling because many of these changes have been observed within the span of one human lifetime, unlike most of the extremely gradual changes that our planet has gone through in its 4.5 billion–year history.

Looking at satellite photos of the Arctic polar ice cap taken over the last thirty years, it's easy to note a dramatic change in its size in a time span of less than one generation. The Arctic and Antarctic regions are the most sensitive areas on Earth to changes in global temperature, in part because the difference in temperature between water in ice form and liquid form is relatively small.

Some people acknowledge that global warming is a reality but try to argue that it isn't causing a climate crisis at all; they point out the potential benefits from some of the effects of a warmer global temperature, such as warmer winters in the more moderate latitudes. Another potential benefit often mentioned is that a decrease in Arctic ice would provide a better northwest passage that could be used for global transportation, making it easier for ships to travel from one side of the planet to the other by using a route north of Canada.

Indeed, these may well be isolated benefits that might occur as the observed trends continue. However, there are many more disadvantages to global warming than there are benefits. In fact, the loss of ice on our planet means there will be less sunlight reflected back into space. White snow and ice, such as that on Earth's poles and glaciers, is much more reflective than the blue ocean or most of the land masses. This reflective property is referred to as *albedo*; the higher the albedo, the more reflective the planet's surface is.

As Earth loses some of its ice to global warming, the planet's albedo will decrease. More sunlight will be absorbed rather than reflected back into space, and this will lead to greater amounts of infrared radiation produced.

The heat that results will of course remain trapped within the insulating effects of the greenhouse gases in our atmosphere and contribute to even greater temperature increases. The entire process has the very real possibility of becoming a runaway reaction, feeding into itself with a never-ending cycle of increasing temperatures, melting ice, decreasing albedo, increased absorption of sunlight, increased infrared radiation, increased global temperatures, and so on. I think the devastating effects that could result from this process far outweigh the benefit of easier sea travel through the Arctic or a milder winter in Canada. (And I don't believe I'm the only one who thinks this way.)

Obviously, polar ice caps aren't the only source of ice on our planet that can melt. Another loss of ice from global warming is the retreat of the planet's many glaciers. Many dramatic examples are available to demonstrate this, but I'll mention only one because the point doesn't need any greater emphasis in my opinion than that. (Many glaciers are retreating, however, and you would be hard-pressed to find one that has been expanding in size in recent years.) The Grinnell Glacier is located in Glacier National Park in Montana. Photographs taken over the years from the 1930s to the present day have shown an obvious retreat in its size, with very dramatic changes that have been easily observed in such a short time. Glaciologists predict that there will be no glaciers remaining in the park after 2030 based on the current trends in global warming that have been observed. When that happens, it seems likely that this once-beautiful natural wonder and the national park where it's located will both need a name change. (Admittedly, such name changes would be the least of the problems associated with melting glaciers.)

With all this ice melting, the added water has to go somewhere; what impact it has will depend on where the ice was located in the first place. For example, Arctic ice is floating freely in the ocean like a giant ice cube because, unlike the Antarctic, there isn't a land mass at the North Pole. When ice at the North Pole melts, it becomes liquid water. What may not be immediately obvious, however, is that water levels in the ocean won't rise as a result of this ice melting. When floating ice melts, the water level doesn't change.

You can easily prove this to yourself by doing a simple home experiment. Take a large clear glass and fill it halfway with water. Make a mark on the side of the glass where the water level is located with a marker or wax crayon. Then add one or two ice cubes so that they can float freely in the glass—make sure they aren't stacked on top of each other or wedged into the glass. The water level will climb on the side of the glass because the ice cubes will displace their weight in water. (The definition of anything that floats is that the volume of water it displaces will weigh less than the volume of water equal to its mass. A simpler way of stating this is that whatever floats is less dense than the liquid it's floating in.) Make a second mark on the glass to show this new waterline. Now leave the glass alone and let the ice cubes melt. When you check the glass a few hours later and the ice cubes have all melted, where is the waterline? You might be surprised to see that it's at precisely the same level it was after the ice cubes were first added to the glass. As they melt, they add their mass in water to the glass, equal to the amount they were displacing earlier while they were floating ice cubes, so the level doesn't change. (This is why the ice cubes had to be freely floating and not stacked or wedged into the glass. Otherwise, they would be displacing only their volume, not their equal mass of water, and the water level would actually drop as they melt.)

What does this simple experiment tell us about the melting of Arctic ice? The ice at the North Pole is really just a large floating ice cube with some smaller ice cubes around its periphery. When this ice melts, the amount of water added is the same as the amount of water it was originally displacing; therefore, the overall water level on the planet won't change.

Then where does all of the concern over rising sea levels come from? The problem is that Arctic ice isn't the only ice on our planet that's melting. There's plenty of ice on land masses such as Antarctica and Greenland, with large ice shelves hanging over the water, some more than a kilometre thick. There are also a number of glaciers just like the Grinnell Glacier all over our planet. Since these sources of ice are on land rather than floating, they aren't displacing any water, a situation that's different from the ice in the Arctic. As glaciers melt, or as large masses of ice break off from ice shelves and make their way into the ocean, they are indeed adding to the global pool of water, making the water levels rise. Land ice melting is akin to adding ice cubes to the glass in the first part of our experiment and noting that the water levels rose. It doesn't matter whether the addition takes the form of ice (before melting) or water (after melting). It still leads to a rise in sea levels.

These processes generally do make our ocean levels rise, and that fact has been verified. According to Australia's national science agency (the Commonwealth Scientific and Industrial Research Organisation, or CSIRO), the current rate of rise for ocean levels is close to 3.0 millimetres, or a little more than a tenth of an inch, a year, proven by careful observations using satellite technology. Earlier in the century, it was closer to about 1.8 millimetres, or 0.07 inches, per year. Therefore, the *rate* of rise is actually increasing, as one might expect given the trends of increasing greenhouse gas emissions and the rising global temperatures associated with them. In addition to the added pool of water from melting ice, part of the rise in ocean levels is also due to expansion of the ocean itself, resulting directly from the increase in temperature because warmer water is less dense than colder water. In one century, that could mean an increase of twenty-six centimetres or ten inches if this rate of rise remains constant, and even more if the rate continues accelerating exponentially, as it has been.

As ocean levels rise, water levels will increase along the world's shores, leading to coastal flooding. Coastal flooding will have major consequences because much of Earth's population is located near sea level. About two-thirds of the planet's cities with a population greater than 5 million people are vulnerable to a rise in sea level. In China, for example, 11 percent of the population (144 million people) will be adversely affected by the rise in sea levels predicted for the next century. The change is slow, unlike a tsunami or hurricane with the rapid flooding associated with those disasters, but it will still have a significant impact on the many millions of people who will be living there.

Another consequence of greater levels of greenhouse gases, including carbon dioxide, in the atmosphere will be greater amounts of carbon dioxide absorbed into our oceans. Some skeptics consider this to be one of Earth's natural compensatory mechanisms to deal with the problem so that we don't have to, but it's not as simple as that. Since carbon dioxide is a gas, it can easily become absorbed into water just like oxygen can. (Otherwise, fish wouldn't do too well—they need oxygen just like we do, and they filter oxygen in the water through their gills.) Recall from chapter 2 that once absorbed into water, carbon dioxide can form carbonic acid, commonly found in carbonated beverages.

$$CO_2 + H_2O \Leftrightarrow H_2CO_3$$

Since the back-and-forth arrow separating the two sides of the equation is present, this is an equation showing equilibrium. That means that if everything is in a closed system with nothing added or removed, then the amounts of the carbon dioxide, water, and carbonic acid will remain constant for a given temperature. Adding more carbon dioxide, however, will drive the equation to the right, producing more carbonic acid; conversely, if carbon dioxide has a chance to escape and vent away, then the equation will be driven more to the left, like a carbonated beverage going flat.

Thus, more carbon dioxide being added to the atmosphere means more is absorbed into our oceans, producing more carbonic acid in the process. This will lower the pH in our oceans. The pH is a measure of how acidic something is. (The term pH likely refers to the "power of hydrogen," although there is some debate as to how the term truly originated.) Something neutral, such as distilled water, has a pH of 7. Anything with a pH less than 7 is acidic, and anything with pH greater than 7 is basic. (Basicity is the opposite of acidity, just as a base is the opposite of an acid. When an acid and a base combine in chemical reactions, they produce a salt and water. It always comes back to some chemistry, doesn't it?)

The mathematics involved are too complicated to review here, but it's worth knowing that the pH scale is logarithmic, which means every number lower on the scale indicates that a substance is ten times more acidic. As an example, orange juice which contains citric acid has a pH of around 3, and the acid in your stomach known as hydrochloric acid is closer to 1. Since gastric acid is two lower than orange juice on the pH scale, stomach acid is about 100 times more acidic because it is two factors of 10 away, and 10 × 10 = 100. In the other direction, bases include baking soda, which has a pH of around 9, and bleach has a pH of about 13.

The pH of our oceans is about 8.2 at present, but with the amount of carbon dioxide being added and more carbonic acid being produced as a result, the National Oceanic and Atmospheric Administration in the United States projects that the pH in our oceans will fall to around 7.8 by the beginning of the twenty-second century. That may not seem like much, but since it's on a logarithmic scale, it's a lot more of a change than it appears. It means our oceans will have increased their acidity by about 150 percent since the start of the Industrial Revolution.

The impacts of such a change in ocean chemistry can be devastating.

Species that build shells, such as crabs and lobsters, won't thrive very well because the calcium carbonate incorporated into their shells will have a greater tendency to dissolve in such an acidic environment, threatening these species with extinction. Coral is another sea-dwelling organism that requires calcium carbonate to exist, so a more acidic ocean will slowly dissolve our coral reefs as well. This not only threatens the coral directly but also threatens the many species that cohabit with them in their complex ecosystems, including more than four thousand species of fish along with many mollusks and crustaceans, all of which use coral as places to live and breed. Corals contribute important medical benefits to our species as well; they provide compounds that have been used in medicines for cancer and HIV/AIDS and have also been used in bone grafting for humans. A more acidic ocean also will likely interfere with the healthy maturation of some fish larvae, possibly threatening them with extinction.

In other words, this small change in the ocean's pH may have consequences to our planet that are almost too difficult to imagine, but given how much we rely on the animals in the ocean for food, we may be dooming ourselves to negative health and economic effects if we allow these trends to continue.

Severe Weather

Another consequence of global warming is the potential for greater storms. Let's look at hurricanes as one example, although entire books are devoted to the changes in weather and ocean currents resulting from rising temperatures. Hurricanes require two things: heat and moisture. Rising global temperatures obviously provide the extra heat in this equation. As for moisture? In the last chapter, we saw that the warmer the air is, the more moisture in the form of water vapour it can hold. So with these two simple criteria met as a result of the documented rise in global temperatures over the last century and a half, it should be no surprise that our planet is experiencing more intense hurricanes that are causing greater amounts of damage than in previous years. One only needs to look at the devastation from Hurricane Katrina in August 2005 to see the extent of havoc that can be caused. That particular storm was the most expensive in US history, costing more than $80 billion and leading to the loss of 1,800 lives. It was the sixth-strongest Atlantic hurricane ever observed. With the continued rise in global temperatures, storms like Katrina will become the rule rather than the exception in the future.

A hurricane as seen from space.

Temperature records are consistently being broken in the modern era, giving further evidence of the effects caused by greenhouse gas emissions, just as the science would predict. The year 2009 was the second hottest on record, running just a fraction of a degree cooler than 2005, and 2010 managed to tie the record. The decade from January 2000 to December 2009 was the hottest decade in human history, and that's not surprising because the years 2002, 2003, 2005, 2006, 2007, and 2009 all made the top-ten list for the hottest years ever recorded.

There are skeptics who try to discredit these observations, but they generally tend to use data from only a small portion of the globe when they make their arguments. For example, Russia, which is the world's largest country, occupies only about 11 percent of the global land mass. The next three largest countries on the list are China, Canada, and the United States, all of which have between 6 and 7 percent of the global land mass each. Thus, it's easy to see how observations that are restricted to only one nation, or more typically, only one part of one nation, will not accurately reflect what's happening on a more global scale. Setting aside the concerns of some skeptics about what is happening in their own backyard, global temperatures are definitely on the rise, just as one would predict based on our understanding of the science associated with greenhouse gas emissions.

A common example of the incomplete evidence used against global warming is the trend that has been noted in recent years of increasingly harsh winter storms. It seems counterintuitive to what you might expect—that global warming would be associated with more severe winter storms—based on the facts already described, and it's a favourite argument of skeptics who claim that the climate crisis is a myth. It seems that every year, some particularly harsh winter storm arises somewhere close to home. The Eastern Seaboard in North America is particularly prone to such storms. Every time this happens, someone uses the storm to argue that global warming isn't real. But the irony is that these storms actually *do* support the evidence of global warming and wouldn't be occurring if not for the fact that global temperatures are on the rise.

You are already aware that warmer global temperatures lead to greater moisture content in the air. However, air masses don't stay in the same place; they move constantly. Sometimes they move into regions of the planet that are colder, such as more northern latitudes. Once they've cooled off, these air masses are no longer able to hold onto all of the water vapour they contain, and the water molecules have to precipitate out. If it's above freezing, then the water will still precipitate as rain, but if the region is cold enough, then it will precipitate as snow instead. If there is greater moisture content in the air due to increasing global temperatures, then more snow will fall from these air masses, resulting in more intense winter storms. Until the planet warms up enough to make snowfall a thing of the past, we can expect to see ever more violent snowstorms in the winters, just like there are more violent hurricanes in the summers. In fact, global warming leads to a significant change in the distribution of water on the planet overall; some areas will experience droughts, and some will experience floods. All of these severe weather phenomena that are being observed today fit with the current understanding of the climate crisis contributed to by our emissions.

Earlier, I described the melting of the polar ice cap as well as the melting of Earth's glaciers. I pointed out that sea levels wouldn't rise from the melting of Arctic ice since it's really just a large floating ice cube. Other than the obvious catastrophic effects that the loss of Arctic ice might have—drowning polar bears come to mind and are commonly used in campaigns aimed at fighting global warming—you might not immediately think of some of the other types of devastation it could cause. But this melting Arctic ice has the potential to affect weather in another manner as well.

Sea ice generally contains fresh water since the salt is essentially expelled from it while it is forming, a process known as *brine rejection*. As ice melts at the North Pole, it adds a large source of fresh water into the surrounding ocean. Some scientists worry that this has the possibility to slow down the Gulf Stream in the Atlantic Ocean, perhaps even stopping it altogether if the problem becomes severe enough. The Gulf Stream carries warm salt water from the western Atlantic Ocean near the Caribbean toward the northwestern regions of Europe. The Gulf Stream gives up its heat there—the main reason palm trees grow in Cornwall at the southwestern tip of Great Britain. Once the heat has been released, the colder and therefore denser salt water sinks down deeper, returning back toward the Caribbean. It's all part of a large ocean conveyor belt that exists all over the planet, and it plays a significant role in our planet's climate and weather.

However, the addition of significant amounts of fresh water from melting Arctic ice carries the risk of slowing down this global conveyor belt of heat within our planet's oceans. There is already evidence that the Gulf Stream runs more slowly than it did in 1950 by as much as 20 percent or possibly even more. If this trend continues, it will affect the Earth's weather patterns severely, with Great Britain and Scandinavia experiencing much colder winters and summers than they have previously, even though much of the planet is breaking temperature records for hotter summers and warmer winters at the same time.

Even though global temperatures are rising, this is just one of the many examples of how a maldistribution of that heat energy can lead to greater extremes across the board. Some regions on Earth will experience hotter weather, and others will be colder. Some will have harsher snowstorms and greater rainfall, whereas others will be exposed to greater dry spells and droughts. Whatever the different regions of the world might experience, conditions will continue to become more extreme than they were prior to the Industrial Revolution.

Spread of Disease

Another consequence of global warming will be the increased spread of disease. Many diseases are passed on from one living organism to another through direct contact. Rabies can come from animal bites, Lyme disease comes from insects, and many bacteria and viruses can thrive almost anywhere, so that simple contact is enough to become infected—one of the reasons we're always being told to wash our hands. The mode of transmission for communicable diseases is known as the *vector*. As global temperatures increase, in many regions, the geographical areas where these vectors live will expand.

Let's look at malaria as one example. Unless you live in or travel to certain specific regions on the planet, it's unlikely that you will personally be at risk to contract the disease, but malaria is a massive problem on a global scale. According to the World Health Organization, about 250 million cases of malaria occur each year, resulting in nearly 1 million deaths among them, most occurring in children under the age of five. Malaria is spread through mosquito bites. You may have heard about the importance of supplying developing nations located in the tropical regions where the disease is endemic with enough mosquito nets for their living quarters to help stop the spread of this disease. The mosquitos don't get the disease themselves, but they carry it from person to person; they are the vector in this case.

In malaria, the name of the bacteria is *Plasmodium*, and there are five different species of it that can infect humans. If a mosquito bites someone infected with malaria, then it will ingest some of the bacteria living in the red blood cells of that infected person. Once a mosquito has ingested these parasites, the parasites will develop inside the mosquito over a week or so without harming it. But once the mosquito bites another victim, the parasites are passed on via the bloodstream. Over weeks to months, the *Plasmodium* bacteria go through their life cycle, initially in the liver of the victim and then ultimately multiplying in the red blood cells, where another mosquito can ingest them, allowing the cycle to continue with the next mosquito bite. Ultimately, malaria can cause fever, hallucinations, and even coma or death.

Obviously, the best way to deal with this disease is by preventing it in the first place—thus the importance of such simple measures as mosquito netting. If people know they will be traveling to parts of the world where malaria is endemic, they can take oral medications to prevent getting the disease in case a mosquito bites them. These preventive medications are known as prophylactic drugs. They are started before the trip so that they're present inside the traveler and ready to disrupt the life cycle of the parasite if he or she is ever exposed from a mosquito bite. Treatment once someone develops malaria involves even more powerful medications than the prophylactic drugs and usually requires hospitalization.

Malaria is generally thought of as a tropical disease. As you can well imagine, when global temperatures rise, the area where the mosquito vector can thrive will expand, increasing slowly into latitudes farther away from the equator, regions of the planet that have always enjoyed more moderate temperatures. When a frost occurs, adult mosquitoes as well as their larvae are killed, but with warmer global temperatures, the frost line will gradually move farther away from the equator. As a result, there won't be as much of a distinction between the tropics and these more moderate latitudes, and the mosquito vector will easily extend its territory and find a new home. The areas where malaria is endemic will expand into regions where malaria has never been considered a health concern, with more cases added to the list and more deaths as a result. And an expanding mosquito territory isn't the only way that global warming will increase the risk of the spread of malaria: hotter weather will lead to longer summers and, therefore, longer mosquito seasons; higher temperatures cause shorter mosquito breeding cycles; female mosquitoes tend to bite more with higher temperatures; and the heavy rains that result from global warming will create more ideal conditions for mosquitoes to breed since they like stagnant water.

Malaria is only one example of the possibility of disease spread due to global warming, but it shouldn't be too difficult to appreciate that the threat of many others exists as well, all as a result of the climate crisis. And these examples aren't restricted only to tropical regions in remote parts of the planet. The Centers for Disease Control and Prevention (CDC) in the United States has confirmed that vector-borne communicable diseases in North America are also changing in both distribution and frequency. Other vectors, such as ticks and mice, are expanding their territories and carrying their diseases into broader territories just like mosquitoes are. Some examples in North America include Lyme disease, Rocky Mountain spotted fever, and certain types of encephalitis, a serious and potentially disabling or even fatal viral infection of the brain. And since global transport is so rapid today, with a flight to Europe or Africa taking only a matter of hours, diseases that were previously considered endemic to faraway regions of the world, such as West Nile virus, can truly become global problems in the present day all too easily.

All the consequences of global warming outlined in this book—melting ice, increasing ocean acidity, rising sea levels, severe weather, and disease transmission—lead to another important though indirect consequence. The costs incurred by the planet's people and nations in dealing with all of these devastating effects are not trivial. The exorbitant costs associated with hurricanes, severe snowstorms, droughts, and coastal flooding are beyond comprehension. Taking into account the increasing risk of disease from expanding vectors of transmission, and deaths from both disease and extreme weather phenomena, the price tag to our civilization is higher still.

Natural Causes of Climate Change

The science supports that the global warming our planet is currently experiencing is our fault: we have been pouring tons of greenhouse gases into the atmosphere for many decades, and we know that the insulating effects they provide will directly lead to an increase in global temperature. However, Earth has experienced many changes in climate in its history since prior to the Industrial Revolution, and some people question whether the changes we're observing might be of those type. In other words, could the natural causes of climate change apart from human activities be playing any part of the problem? Certainly the skeptics make this argument frequently when they try to argue that we shouldn't bother to make any effort to curb our emissions. The short answer is that none of the natural causes of climate change are actively contributing to what we're experiencing, but it is worth understanding some of the basic concepts behind these natural processes that have affected our planet in the past and will again in the future.

The easiest approach to understanding this is to think in terms of time scale. There are natural processes that affect our planet's climate over decades, centuries, tens of thousands of years, and millions of years. More simply stated, these can be thought of as immediate, short-, medium-, and long-term effects.

The immediate effects on our planet generally result from internal variations within our climate. These include natural fluctuations that occur because of the dynamic interaction between land masses and the oceans surrounding them. Two of these are well known, both existing as natural variations known as oscillations. One is the El Niño–Southern Oscillation in the Pacific Ocean, and another is the North Atlantic Oscillation. They affect climate over decades, but these effects are regional rather than global and do not explain changing trends in worldwide temperatures like those we've been witnessing, so they aren't a possible alternate explanation to global warming.

The short-term effects on climate result from changes in the sun's amount of *irradiance*, or more simply the amount of sunlight it produces. The sun has a naturally occurring cycle of eleven years, with changes in sunspot activity and minuscule changes in irradiance, but these have had no more than an extremely minor impact on the amount of sunlight that Earth has been receiving in recent years. However, over centuries there have been more significant changes, and these have likely contributed to some of the known differences in global temperature that have been recorded in history, such as the Medieval Warm Period, which occurred from AD 950 to 1250, followed by the Little Ice Age, which lasted until around the end of the nineteenth century. Since satellites first started recording data about the energy output of the sun around 1978, there have been no significant differences in irradiance observed that would offer an alternate explanation for the changes in climate during that same time frame.

The medium- and long-term natural effects on climate are even less likely to impact on our planet in the ways we're observing because such changes take even longer to occur, more like tens of thousands of years. Medium-term effects result from something called *orbital forcings*. These are changes in the way our planet orbits the sun. There are three main orbital forcings that occur over differing time scales, ultimately contributing to approximately a 100,000-

year cycle. They consist of changes in the Earth's *obliquity*, its *precession*, and its *eccentricity*, terms that sound more complicated than they are.

Obliquity refers to the tilt of the Earth's axis as it orbits the sun, responsible for our four seasons. When Earth's Northern Hemisphere is facing the sun, we enjoy our warmer summers, and six months later, when it's facing away from the sun, we experience colder winters. The Southern Hemisphere experiences the opposite because when one pole is facing away from the sun, the opposite pole faces toward it, and vice versa. Earth's axis is currently 23.4 degrees but varies between 21.5 and 24.5 degrees over a 41,000-year cycle. The greater the tilt, the more extreme the changes in the seasons.

Precession refers to how the obliquity or tilt rotates on its own axis, similar to how a spinning top will slowly change the direction of its tilt, though the actual spin is much faster. This occurs on a 21,000-year cycle and means that Polaris, our current north star, doesn't always have that distinction. In another twelve thousand years, our north star will be Vega in the constellation Lyra. During the transition along the way, there won't be a particularly bright star located at the North Pole; we happen to have one now, but that isn't always the case.

Other than a change in how our night sky is laid out for our view, you might not think that a change in precession would play a very significant role, but it does because Earth's orbit around the sun isn't a perfect circle. It's an ellipse, which is more like a flattened or squeezed circle. The amount of flattening is referred to as *eccentricity*. Our planet's orbit isn't particularly eccentric compared to some other planets such as Mars, but it's enough to have an impact. When Earth is closest to the sun, a point called *perihelion*, it is about 147 million kilometres or 91 million miles away. When Earth is at its farthest point from the sun in its elliptical orbit, known as the *aphelion*, it is about 152 million kilometres or 94.5 million miles away, a variation of almost 4 percent. The perihelion presently occurs in January when the Northern Hemisphere is experiencing winter, and the aphelion is in July during the Northern Hemisphere's summer.

This effect helps to soften the extremes of the seasons somewhat for the majority of the world's population since most people live in the northern latitudes. When it's the middle of summer for the Northern Hemisphere, we also happen to be the farthest from the sun that we can possibly be. Winters also are a little milder because we are closer to the sun during the middle of that season. Things will be different in another twelve thousand years, however, because both the precession and the eccentricity of Earth's orbit will reinforce each other in the Northern Hemisphere, making for a greater variation in temperature from summer to winter than we experience now. One thing is clear, however: orbital forcings take far too long to be contributing to what has been taking place since the Industrial Revolution began.

The long-term natural impacts on climate over millions of years tend to exist because of various changes that understandably take a very long time to occur. A significant one referred to previously in chapter 2 is tectonic-plate movement. When land masses change and create mountain ranges in the process, for example, this has an impact on climate. It's easy to appreciate that such changes won't occur overnight, however, or even over a few hundred years, but they do have an impact given enough time.

In fact, although all of the effects described here have changed our climate in the past and will continue to do so in the future, none of them have had any significant impact in the last few hundred years. The medium- and long-term effects from orbital forcings and tectonic-plate activity take far too long to have any appreciable effect in the brief span of time compared with our increased greenhouse gas emissions. The more immediate effects from oscillations tend to be local rather than global. The only natural effect that could possibly contribute to what we've been observing is the short-term type—namely, changes in solar irradiance—and careful satellite observations have proven that's not the case. We need to look no further than our own standards of living and the greenhouse gas emissions associated with them to explain the changes that we are observing. They are due to the lives we lead; we are the main cause of our present plight, plain and simple.

Armed with this knowledge, it would certainly be prudent if we, the present caretakers of this planet, could do something to try to prevent such catastrophes from reaching a point of crisis with even greater amounts of devastation. But what can we do that will make a difference?

Section 3 addresses some of the solutions that can help combat the climate crisis, along with some of the hurdles we face in adopting them. Chapter 6 addresses some things you can do to reduce your personal carbon footprint, changing what you can to minimize your greenhouse gas emissions so that you and your family can do your part and contribute toward a solution.

Key Concepts of Chapter 5

- Evidence shows that temperatures are increasing in ways consistent with the predictions of the effects of increased greenhouse gas emissions.

- Increasing global temperatures will contribute to many changes in weather and climate, a number of which have already been observed to be taking place.

- Although natural changes in weather and climate exist, none are able to explain the observations made, helping to confirm that human activity is the culprit.

Suggested Reading for Chapter 5

Cox, John D. *Climate Crash: Abrupt Climate Change and What It Means for Our Future.* Washington DC: Joseph Henry Press, 2005.

Gore, Al. *An Inconvenient Truth: The Planetary Emergency of Global Warming and What We Can Do about It.* New York, NY: Rodale, 2006.

Mathez, Edmond A. *Climate Change: The Science of Global Warming and Our Energy Future.* New York, NY: Columbia University Press, 2009.

Mogil, H. Michael. *Extreme Weather: Understanding the Science of Hurricanes, Tornadoes, Floods, Heat Waves, Snow Storms, Global Warming, and Other Atmospheric Disturbances.* New York, NY: Black Dog and Leventhal, 2010.

Section 3
The Solutions

CHAPTER 6

Reducing Greenhouse Gas Emissions: The Small Steps You Can Take

> Never doubt
> that a small group of thoughtful,
> committed citizens can change the world;
> indeed, it's the only thing that ever has.
>
> —Margaret Mead

Every concept or idea always has more than one perspective to it. As a case in point, I was recently involved in a conversation about religion with two close friends. One of my friends was devoutly religious, and the other was emphatically atheistic. Many arguments were made back and forth; I heard about the importance of organized religion and its various benefits to society, but I also heard about the wars and repression over many centuries that have all occurred in the name of a Supreme Being. In the course of this interesting and at times rather tense conversation, one point made by my religious friend was that if he was wrong, he had nothing to lose by living a lifestyle that encouraged everyone to love one's neighbour, to do good wherever possible, and to shun anything that would be considered wrong, whether in the eyes of his friends and family, in the eyes of the law, or in the eyes of his God. But if he was right, then he had everything to gain by living with his choice, primarily consisting of spending a happy, eternal afterlife in paradise, and those who shunned the same choice would have everything to lose, namely an eternity of misery and suffering.

Now I'm not sure that he made a particularly good argument with that point because I would figure that a Supreme Being would know whether someone was a true believer or simply hedging his bets. Religious beliefs notwithstanding, however, I did think that his argument could be applied to the choices we all have to make with respect to trying to preserve our environment and protect the climate by reducing our personal carbon footprints and doing what we can do to cut back on our greenhouse gas emissions. If we are wrong and these measures aren't as critical as most scientists believe they are, then what will we have gained? A reduction in pollution, a more resourceful and less wasteful society, and a sustainable solution in terms of thriving long after fossil fuels are depleted, something we as a society will have to face eventually anyway. But if we are right about the very pressing need to change, then we have everything to gain, perhaps nothing less significant than the long-term health of our planet and all life on it. To me, it seems like a win-win situation, and this is one of the reasons I think we should try to do whatever we can, whatever is within our means, to help reduce our emissions. In other words, I think it's totally reasonable to hedge our bets on this one.

Why We Need to Do Something

To those who choose not to heed the facts on the climate crisis and its consequences, despite what almost every respected scientist has had to say on the issue, there is generally nothing that can be added to the list of arguments to convince them further. They have made up their minds. But I believe that ignoring the issue is a huge mistake. If the perspective of the skeptics is indeed wrong, and we've done nothing to help, we have everything to lose. More accurately, our children, our grandchildren, and all future generations have everything to lose, facing the dire consequences of our inactions. And unlike an experiment that has gone wrong, there is no way to reset it, looking for a better outcome the second time around. This is the only planet we have. If we're wrong in continuing on with business as usual, we'll have no "do over" available to us. Dr. Roger Revelle, who along with Dr. Charles David Keeling made the observations about the annual fluctuations in carbon dioxide levels described in chapter 4, said it particularly well: "Human beings are now carrying out a large-scale geophysical experiment of a kind that could not have happened in the past nor be reproduced in the future" (Revelle, 18–27).

Section 1 of this book presented factual explanations about how Earth and all of the elements in it came to exist and how carbon in particular has been circulating throughout our planet in something known as the carbon cycle; some of the pathways in this cycle led to the creation of fossil fuels. Section 2 addressed how we as a species have made significant changes to the atmosphere and our climate by the burning of those fossil fuels and through agriculture and deforestation, along with the consequences of those changes.

The majority of human activities in the twenty-first century are designed to make our standard of living as high and as convenient as possible, to levels never previously seen for many on the planet. Unfortunately, these same activities put millions of tons of carbon dioxide into the atmosphere each year, a known greenhouse gas with the unattractive side effect of increasing our global temperature. This adversely affects our weather and our climate, leading to all of the devastating effects those changes can cause. The vast majority of scientists who have studied these issues have agreed that this is something real and that it's a crisis we have to address before it's too late. Many believe we have already passed the tipping point, but a number believe we have some precious time left to prevent very serious long-term harm to our planet, as long as major efforts are made to tackle these problems and soon.

Interestingly, if there are differences of opinion among scientists on the issue of the climate crisis, they are not on the question of whether the crisis exists but rather what the crisis timeline is and whether we will face the tipping point in forty years, twenty years, or ten years, or whether that point has already come and gone. For those who argue that we are very close or have already passed it, there is a real sense of grief and frustration because we aren't witnessing the rate of change toward a solution that we need to.

Research led by Dr. Nathan Gillett at the Canadian Centre for Climate Modeling and Analysis of Environment Canada made projections for the next thousand years based on information about where we are at present. His group's research suggests that even if we dropped our emissions to zero right now, higher temperatures would dog us for the next millennium. Given how long carbon dioxide stays in our atmosphere and how long a warmer ocean holds onto its heat, they don't believe we'll be cooling things down anytime soon. (Think of the analogy of adding blankets to warm up at night—and remember that greenhouse gases are the blankets in our atmosphere: once you're a little too warm, it isn't enough to simply stop adding extra blankets; you have to remove some to get more comfortable.) Unless we find a way to remove carbon dioxide from our atmosphere with some as-yet-undiscovered technology, it may be too late to stop these trends. Until Earth's carbon cycle starts to remove it naturally without our continuing to add more, the changes in climate and global warming that we've witnessed already are likely to continue. But whether we will be able to avert a real crisis situation if we make important strides toward helping the problem is still unknown. Obviously, I believe it's worth the effort.

Skeptics often refer to some of Earth's natural mechanisms to take carbon dioxide out of the atmosphere so that we don't have to go to such Herculean efforts to combat this problem. In the last chapter I referred to how the planet's oceans can serve as a sink for carbon dioxide because they can absorb it, thereby removing it from the atmosphere. But as I also pointed out, the increasing acidity of the oceans associated with that absorption could have devastating consequences for the health of marine life and ultimately us as well.

Some skeptics also point out that trees will naturally respond to the increased levels of carbon dioxide in the atmosphere and ramp up photosynthesis to remove it just as quickly. However, studies show that only about 20 percent of trees on Earth increase photosynthesis as a response to increased levels of carbon dioxide. And regardless of which mechanisms skeptics grasp for to explain why we don't have to make any efforts ourselves, all one has to do is look at the steady climb in atmospheric carbon dioxide levels along with the other greenhouse gases since the Industrial Revolution to realize that whatever methods the planet is trying to use to fight this problem, it isn't doing a good enough job. Since we're causing this problem, it only makes sense that we should try to fix it as well. As this chapter continues, I will address some of the possible solutions to the problem as well as some of the hurdles we have to overcome.

WHAT WE CAN DO TO HELP

For those readers who are concerned enough about the environment and want to help (and I would hope that is the vast majority if not all of you), this chapter addresses what you as an individual can do to contribute toward a solution, from the simple to the substantial. Every little bit helps, and if everyone made some changes, a significant reduction in greenhouse gas emissions would occur. Chapter 7 looks at larger-scale solutions, those that societies, countries, and governments can consider on a national or even global scale, but also at some of the hurdles standing in the way of our making such changes as quickly as we might like. Chapter 8 tackles the problem of changing attitudes and why society can be so resistant.

But this chapter is just for you, dear reader. There are only two things standing in the way of people making these changes. The first roadblock is often a lack of motivation to change a fixed routine. The second barrier, especially for some of the more ambitious changes, is having adequate funds to afford them. I can guarantee, however, that there are changes listed here that everyone can and should make.

The main categories of changes that individuals can undertake fall under the following:

- Improved efficiency and reduced consumption
- Energy-efficient vehicles
- Carbon offsets
- Renewable energy sources

MORE EFFICIENCY, LESS CONSUMPTION

This is the simplest way to make a personal change, and unlike the other categories, improving energy efficiency in your daily life doesn't have to cost much. In fact, depending on how inefficient you currently are, it can actually save you a lot of money. You've probably already heard about many of these recommendations; documentaries and books devoted to this very topical subject abound, some of which are listed in the Suggested Reading section at the end of this chapter. Nonetheless, it's worth reviewing some of the basics that many of us haven't considered yet or that we simply take for granted.

- Change all light bulbs to compact fluorescent light bulbs (CFLs); these cost a bit more, but their prolonged lifetime more than makes up for the cost in the long run. They use a fraction of the energy required by their incandescent cousins, invented by Thomas Edison in the 1870s and in use ever since, with only minor modifications in design over the years. We can anticipate even further improvements in lighting in the coming years as light-emitting diodes (LEDs) continue to evolve with ongoing research and development. LEDs tend to use even less energy than CFLs and have the advantage of not containing mercury, something that is present in every fluorescent light bulb and that has its own share of environmental concerns due to the question of how best to dispose of them.

- Purchase the most energy-efficient appliances you can, ones with the Energy Star label. These often cost more up-front but will save you a lot of money on your electricity bill in the long run after years of use. The greater the demand for energy-efficient appliances, the greater the competition will be in their manufacture, and the lower their costs will become. If it's within your means to make these purchases now, strongly consider doing so. Anytime a household appliance needs replacing, make efficiency one of the key criteria in your purchasing decision.

- Have an energy audit performed on your home. For the cost of a few hundred dollars, usually enough correctable deficiencies are identified such that an audit will pay for itself in time. Changes are often such simple measures as improving the caulking on windows, changing the filters in your heating and air conditioning, and installing low-flow shower heads that use less hot water. Often, the government will give you rebates if you make changes based on the audit's results within a certain period of time after the audit was completed.

- Install a programmable thermostat. There's little sense in heating your home to the levels that provide maximal comfort when no one is around or when everyone is asleep and tucked away in cozy beds. The times when we need a warmer house are when we've all gotten up in the morning and when we return home after a day at school or work until the time we go to bed. Although you can manually adjust your thermostat, it's a lot easier to purchase a programmable one that does all of the adjustments for you. You can select the times you want the temperature of the house to change and to what extent. That way, you can have higher and more comfortable temperatures for only a few hours out of a typical day. The house can be much cooler most of the time, but you won't notice because the thermostat program will get the temperature increased to the level you desire before everyone has woken up in the morning. Likewise, it will warm up the house by the time everyone gets home. This not only will save you plenty of money in a year but also will significantly reduce your carbon footprint.

- Reduce, reuse, and recycle. You've heard the phrase many times, but how much do you really put it into practice? Less waste means less manufactured goods; that means less carbon dioxide making its way into the atmosphere as a result of the manufacture and transportation of those goods. Anytime you can go without packaging, do so. Precycling refers to the concept of making smarter purchases that involve less waste, minimizing the need to recycle waste materials later on. From downloading music and movies digitally rather than purchasing CDs and DVDs to bringing your own bags to the grocery store, every way that you can eliminate garbage helps. Billions of plastic bags are thrown out every year, and since plastic comes from petroleum products, this has a significant impact as well. Make sure you separate your garbage into recyclable items, such as paper, plastic, glass, and aluminum. Also separate your food waste as compost. Almost half of food waste still ends up in our garbage. Once it makes its way to landfill sites, it decomposes and releases methane, one of the greenhouse gases we're trying to reduce. If food waste is composted instead, it becomes a material rich in nutrients and useful in soil. You may be able to have your own compost on your property and use it to benefit your garden. However, your community should also have a compost pick-up along with garbage and recyclables. If it doesn't, complain to the city or town council until it does. In the twenty-first century, we have to do what we can to minimize the garbage that's being sent to landfill sites. These are the simple sorts of measures that will help.

- Unplug items from wall outlets when you're not using them. They all draw a small amount of energy even when they're not in use, as much as 10 percent of your electricity bill in fact. If these items have any lights or displays on them, they draw even more energy, and that adds to your electricity usage even further. Unless your electricity is 100 percent green—a topic further explored later in this chapter—this translates into added carbon dioxide that you put into the atmosphere. Newer computers usually have a standby feature that allows the computer to power down substantially if you're not using it for a while; make sure that function is enabled on yours. If you're going to be away from your computer for any extended period of time, it's best to simply turn it off. And when you're away on vacation, there are likely a number of household items, such as televisions, radios, and bathroom and kitchen gadgets, that can be unplugged so as to minimize any unnecessary power drain. This will save you money and reduce your carbon footprint.

- Buy locally and become what is increasingly being referred to as a "locavore." In addition to supporting your local farmers and the local economy, you substantially reduce carbon dioxide emissions because foods that originate a distance away require transportation, and that means more emissions from the trucks that brought them to your grocery store. One simple way to buy locally is at farmers' markets. Another is to look in your grocery store for produce that is grown locally and buy it instead of produce that comes from thousands of miles away. If you can't find it, ask your grocer to get it. If they won't honour your request, change to a grocery store that will. They'll all get the hint with time, and it won't be long before buying locally is the rule rather than the exception. And while we're on the subject of food, you might want to consider reducing some of your meat consumption. Given the significant methane emissions that come from cattle, cutting back on red meat can have a substantial impact. You can make such changes by having "Meatless Mondays" become a part of your weekly routine. Given that most people in North America eat a diet that is much higher in meat—particularly red meat—than it has to be, this can also be a change that will benefit your health as much as it benefits the planet.

- Support businesses that are also making an effort to go green. Consider two frustrating examples where businesses did the right thing, but consumers failed to show support; these cases are disappointing because these companies made efforts to help the problem, but the market complained, forcing them to reverse their decisions. The first example is Loblaws, a grocery store chain in Canada. The stores started to charge five cents per grocery bag as a gentle surcharge to patrons, with the goal that customers would start to bring in their own bags and bins, minimizing the need for more plastic bags to be produced and ultimately disposed of in landfills. Rather than embracing the concept, however, shoppers in Atlantic Canada started to switch to another chain that didn't have an extra charge for bags. What is cheaper is more important than what is greener it seems, and the loss of customers forced Loblaws to remove the charge. The second example is Frito Lay, which invested money into the research and development of a biodegradable bag for its SunChips product line. These would degrade naturally over a couple of weeks, unlike plastic bags that last for years in landfill sites. Unfortunately, sales started to drop because the bags were noisier (which I think is an odd complaint since potato chips are already pretty loud as far as foods go). The company was forced to switch back to the plastic bags for most of its product line in the Unites States, where research continues for a quieter version.

(Interestingly, in Canada, Frito Lay has taken a different stand and continues to use the noisier but greener version. They've even used it as a marketing gimmick, encouraging their customers to send away for their SunChips ear plugs and asking them to write down why they're happy to make some noise for the environment.) These two examples demonstrate that although many in society want to go green, there are a few things they aren't prepared to give up. If you believe in the greener efforts that businesses are making, support them any way you can.

- Downsize your life. Our society focuses so much on material items that we tend to forget some of the more important components of a happy and healthy life. We are overly obsessed with more toys, bigger cars, the latest electronic gadgets, and huge homes. We can have less of many of these items and still lead as productive and fulfilling a life as we do now. Think of your house and how many rooms you use only rarely or not at all. Even if you close the vents in those rooms, you're still paying to heat them to some extent, adding greenhouse gas emissions in the process. If we could all live a little more modestly by reducing our living space in general, we'd substantially reduce our carbon footprint. If you're looking for a house, make sure it's no bigger than what you truly need. The British economist E. F. Schumacher put it both eloquently and succinctly with the title of a collection of essays he first published in 1973: *Small Is Beautiful*.

ENERGY-EFFICIENT VEHICLES

The vehicles we drive account for about a third of our carbon dioxide emissions. If we can reduce those emissions, we can make a substantial impact on our carbon footprint. Some of us can pursue such changes as biking to work or carpooling, but not everyone has that option, and almost nobody has that option all of the time, even beyond work. There are occasions where most of us will require a vehicle to drive, and no other option will do. Despite that fact, there are many unnecessary trips in cars that we can likely do without, and we definitely need to make efforts to reduce our fuel consumption. Along with the benefits to the climate, energy-efficient vehicles have the added bonus of reducing pollution with the associated benefits to everyone's health. Here are some things to consider:

- Your existing vehicle can often have its efficiency improved with such simple measures as getting a tune-up, using the highest grade of octane you can afford, and ensuring your tires are inflated to the proper amount.

- When you're looking to replace a vehicle, consider purchasing the smallest vehicle that you need. Smaller means lighter and thus the burning of less fuel to move it from point A to point B. Similar to the recommendation to use efficiency as one of the criteria in deciding what to buy when replacing a household appliance, look into the energy efficiency of the vehicles you're considering. Hybrid vehicles have the advantage of charging the battery during braking and coasting—known as *regenerative braking*—which is then used to help power the vehicle, so that overall, less fuel is consumed. They generally cost more than conventional automobiles and may not necessarily pay themselves off in the long run, depending on your driving habits, but they use less fuel than a comparably sized vehicle that solely uses gasoline or diesel. In the future, we can look to completely electric vehicles as well. Some point out that if electricity is used to power our automobiles, and the burning of fossil fuels such as coal generates that electricity, then they aren't really helping the problem since the greenhouse gas emissions will still occur from the generation of the electricity in the first place. But if the electricity used to charge them is green and not from fossil fuels, then these vehicles will significantly reduce our emissions.

PURCHASE CARBON OFFSETS

Presently, it's easier than ever before for people to learn how to make a difference because information has never been more accessible than it is today. You can easily do an Internet search for carbon footprint calculators and figure out approximately how much you and your family are contributing to the problem. To see the numbers come back at you is usually an eye-opening experience. It would not be unusual for a typical family to put about 25 kg of carbon dioxide into the atmosphere each and every day, with about 60 percent of that coming from household electricity and natural gas use and the other 40 percent coming from transportation.

Obviously, this can vary dramatically from one person to the next and one family to the next. For example, someone who walks to work will contribute much less than someone who has a one-hour commute on a daily basis. In turn, the commuter will contribute much less than the executive of a multinational corporation who racks up air miles and flies all over the world on a weekly basis. Having an idea of just how much carbon dioxide you and your family contribute is one of the simplest ways to motivate yourself to make a change. Online calculators will often let you see what the difference will be when you make some of the changes described previously, encouraging you to make that next automobile or appliance purchase an energy-efficient one. But in the meantime, we all still contribute to the problem to some extent as we continue to live our lives.

Since almost none of us can reduce our emissions to zero, one way to help mitigate our contribution is to purchase *carbon offsets*. Various reputable organizations sell these offsets for your personal use, and I encourage you to explore these options. Depending on how much carbon dioxide you and your family add to the atmosphere, which you can figure out with the online calculators, you can purchase these and, in so doing, help to counteract—or offset—your contribution. This is typically done on a monthly basis, similar to another household bill. The money you spend is used to aid in the development of renewable energy sources, such as solar, geothermal, and wind farms. Often, these developments are going to be in countries far away from where you live, but since this is a global problem—your personal emissions don't stay in your neck of the woods but end up scattering across the entire planet's atmosphere over hundreds of years—there's nothing wrong with contributing in this way toward a global solution. As an exercise, check out a carbon footprint calculator online and then find out what your carbon offsets would cost. In our house, we purchase carbon offsets to cover the cost of our automobiles and the driving we do. Our monthly cost is about thirty-five dollars, something we are more than happy to pay. You can choose to offset your cars, your home, your plane travel, or any combination or even all of them, depending on how motivated you are and how much you're willing to spend to help.

Many groups and organizations are making changes by purchasing carbon offsets as well. Some musicians are doing it to help negate the effects of the manufacture of their CDs and the fuel they use while travelling on tour. Politicians and corporations often do it to offset the carbon footprint associated with their travel needs. In addition to purchasing offsets on a monthly basis to counteract your regular emissions, you can also help with any special occasions that are a departure from your usual routine if you wish, such as your vacations. Most airlines will offer green travel options or carbon offsets for any given flight, so you can purchase these along with your flights to make sure your next trip is a green one.

Renewable Energy Sources

Eventually, given enough time, this planet will run out of fossil fuels. The sources of oil in the Middle East have been providing much of the world's petroleum products for many decades, but those sources will become depleted one day. For everyone to have access to fossil fuels, other sources need to be developed. Some of these include offshore drilling—looking for untapped pockets of oil beneath the ocean floor—and oil sands, such as those in Alberta, Canada. Overall, these other sources are not as efficient as the desert sources on which we have predominantly been relying so far. That's because more energy is required to get a barrel of oil out of an offshore site or from the oil sands than from a desert in the Middle East, but since the global economy is wrapped up in oil and fossil fuels, they're still generally being pursued because they remain profitable business ventures. Because these sources are trickier to obtain oil from, however, they also tend to be more harmful to their local environments.

Still, despite the exploration of these other sources, eventually they will run out as well. Since fossil fuels take millions of years to produce, they will never become replenished at a rate comparable to the world's consumption. Something else needs to come along to replace them; no matter how we look at it, the simple fact is that we as a society would have to look to alternate sources of energy eventually even if climate weren't a motivating concern. The best sources are those that will continue to be replenished at a rate that keeps up with our consumption, so-called renewable sources. These consist of solar, wind, and geothermal.

Solar

Solar energy taps directly into the obvious source of most of Earth's energy: the sun. In fact, the amount of energy that reaches the surface of the Earth is almost too vast to comprehend. Our planet's surface receives enough solar energy in one hour to match the energy needs of the global population in one year. Enough reaches Earth in one year to equal more than double all of the energy we have derived from all fossil fuels and nuclear energy combined—throughout all time! Of course, a lot of sunlight reflects off the clouds and oceans and therefore isn't so easily accessible as an energy source, but the point is clear: there's plenty of potential energy available to us through the use of the power of sunlight. In solar power, we have a significant amount of energy that isn't going to run out anytime soon—for a few billion years at least—and it will be clean in the sense that it won't produce any emissions in its consumption.

The best way for people to use solar energy is by installing solar panels. These are called *photovoltaic panels*, and they use a principle known as the photovoltaic effect. You will recall that in chapter 4, I described visible light as part of the electromagnetic (EM) spectrum. I described light energy as waves, but I also pointed out that light behaves as particles with the name of photons. Thinking of radiation in the EM spectrum as waves works best for the purposes of understanding how visible light is absorbed by the Earth's surface and infrared radiation is reflected back as heat. When you're thinking about light hitting photovoltaic panels to produce solar energy, however, it's best to think of light as photons. These little particles of energy can knock electrons out of their orbits around their nuclei. This simple process creates a *direct current* (DC) that can be used as a source of power to something it's connected to nearby, that can be stored in a battery, or that through an inverter can be changed into an *alternating current* (AC), which is the electricity we use in our homes. (Direct current can't travel through electrical wiring more than one to two kilometres or about a mile before power starts to be lost. In contrast, alternating current can travel hundreds of kilometres or miles without losing power. That's why it's AC electricity we get from the outlets in our homes. DC power is most commonly provided by batteries in low-voltage items, such as toys and radios.)

People can buy solar panels and install them on the sides of their homes or up on the tops of their roofs. Depending on how many panels they install, and how much sunlight exposure they receive at their particular climate and latitude, they can aim to produce enough energy to provide the electricity for their entire home or for just some small portion of it, such as the hot water heater. Some people don't like the appearance of solar panels, and the costs involved are often too prohibitive for many to even consider them an option, but it's highly likely that they'll become an ever-increasing presence in homes and commercial buildings in the years to come.

Wind

The heat we get from the sun also heats up the atmosphere, and this causes air currents to move. That won't end anytime soon either, so if we can harness the energy from the movement of wind, we will have another source of energy that is easily replenished and clean. The oldest method of trying to harness wind power for any industrial purpose is the windmill, and although it has changed somewhat over the years, its basic principle is the same. With windmills, wind turns the sails, converting wind energy into a rotational motion that is used to power a mill. (Did you ever think about why this structure was called a windmill? It used wind energy to power a grinding mill or paper mill, although windmills often have been used for other purposes, such as pumping water.)

In the modern day, windmills have been replaced with wind turbines. (Turbines and how they generate electricity because something rotates them were described in chapter 3.) Wind turbines use the power of the wind to create rotational energy similar to windmills, but instead of using that energy directly to grind flour or pump water, they rotate a turbine and generate electricity. Turbines are the main method for creating energy; coal, hydro, and nuclear energy all use them to generate electricity, with coal and nuclear producing steam to turn the turbines and hydro using water. In the case of a wind turbine, wind generates the rotational energy; wind is clean and renewable, unlike fossil fuels.

Most people don't have the kind of home or property that would lend itself well to a wind turbine, but a few out there do. As with solar panels, issues of appearance and cost often prohibit people from considering it as a viable option.

Geoexchange

Geothermal energy uses the heat within the ground as a source of energy. More properly, geothermal energy refers to the heat residing deep within the Earth, often kilometres or miles down. This is discussed more specifically in the next chapter because it's something used on a much grander industrial scale rather than as a direct source of energy for individuals' homes. The process addressed here is better referred to as *geoexchange*, also known as a ground source heat pump. Despite the dramatic fluctuations in the surface temperature throughout the year from summer to winter, once you dig down two metres or about six feet into the ground, you have a constant temperature of about 10 degrees Celsius or 50 degrees Fahrenheit in the more northern latitudes and generally warmer the closer to the equator that you live. This constant temperature can serve as a heat source, providing heat for your home in the winter and a heat sink to help cool your home in the summer. All of this is done with a significant reduction in electricity consumption and carbon dioxide emissions. The system uses the same principle as your refrigerator, where a liquid coolant circulates through a series of coils.

The process of refrigeration is a bit complicated but involves compression followed by evaporation of a circulating liquid, often Freon. During this process, the heat from inside the refrigerator can be transferred elsewhere, namely outside. Since everything in a room gradually comes toward the same ambient temperature naturally, with hot items cooling down and cold things warming up, this process of keeping the inside of a refrigerator cold and a freezer even colder takes energy. Your refrigerator is plugged in, and electricity provides the energy for the motor, fans, and condensers. If you explore the surroundings of your refrigerator, you'll feel the heat either below or behind it. Part of that is the heat from the motor, but part is the heat that was removed from the refrigerator's interior.

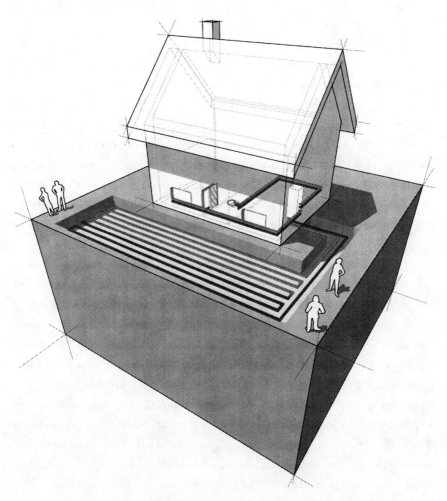

A system of geoexchange used to heat and air-condition a home.

With geoexchange for your home, a series of pipes are dug into the ground below the frost line where the temperature is constant year-round. In the winter months when you need to heat your home, the cooler pipes circulating under the surface will absorb the heat from the ground. That heat will be returned to the interior of your home and will then be subjected to a compression/evaporation process similar to that in your refrigerator. In this way, the heat is collected and then pumped throughout your house with a constant flow of air, different from the cycling that occurs with a typical furnace. In the summer months, the process is simply reversed, with heat collected from the inside of your house and transferred back to the ground so that cooler air will be pumped instead.

One of the biggest drawbacks of using this type of system in your home is the up-front expense; the system usually takes many years to pay itself off from heating and air-conditioning savings. That's why it's easier to consider geoexchange when a home is being built rather than as a modification to an existing home. Also, electricity is needed for the system to operate. If the electricity used isn't green—and most of the world's electricity comes from coal—then geoexchange furnaces can actually produce more greenhouse gas emissions than a furnace that runs on natural gas since methane is cleaner than coal. But if your current furnace already runs on electricity instead of natural gas, then geoexchange will generally lead to a reduction in emissions.

All of these alternate sources of energy are things that you and your family can consider for your own homes. Solar panels can be applied to your house and provide some or all of the electricity you need, depending on how many panels you put in place, or perhaps just enough to provide all of the hot water you need; a wind turbine can be constructed and generate electricity as well if you are in a location that's windy enough; and you can obtain energy from geoexchange by digging into the ground on your property and burying the pipes that allow heat exchange. The biggest problem with renewable energy sources on such a personal level, however, is the cost; none of these projects are cheap. There may be government subsidies that can help ease the expense, but generally, you're looking at an investment for which it will take many years to make up the cost in electricity savings. For all practical purposes, it might be better to think of these measures as doing something to help save the climate but at a personal cost to you and your family.

Obviously, not everyone is in a position to take on these costly projects. Fortunately, there are still options available to you for using renewable energy without having to shell out thousands of dollars for solar, wind, or geoexchange energy on your own property. One simple way to tap into these resources right now is to ensure that your energy provider gets its electricity from renewable resources. If your electricity provider doesn't, you may be able to switch utility companies to one that does.

One such example is Bullfrog Power in Canada. When people and businesses switch to Bullfrog, which has been an option available since 2005, they still draw from the regular energy grid and still pay to the regular electricity company, but Bullfrog returns the amount of electricity used to the grid. The replaced electricity is much greener, provided exclusively by hydro and wind, bringing emissions down substantially. Even though it costs a few more dollars per month paid to Bullfrog—in addition to the regular electricity bill to the utility company—it's certainly much easier for most people to afford compared with the substantial costs of providing their own solar, wind, or geoexchange energy.

In 2011, Bullfrog added green natural gas as an option as well. Here, the natural gas associated with decay in landfills is captured, cleaned, and then injected into the pipeline network. The difference is that this natural gas is part of a natural cycle and would make its way into the atmosphere one way or the other anyway, unlike the natural gas that's sequestered inside the Earth until we drill for it. Just like with the electricity option, people and businesses continue to draw from the utility company and pay a bill to it, but Bullfrog replaces the amount of natural gas used.

Still, not everyone is prepared to pay even a little bit extra to help improve an individual carbon footprint, but for those who are motivated and consider the slight increase in cost worthwhile, obtaining electricity and natural gas from renewable resources is a huge move in the right direction that almost everyone can do. As a case in point, as of 2011, Bullfrog is available to Canadians living in the provinces of British Columbia, Alberta, Saskatchewan, Manitoba, Ontario, New Brunswick, Nova Scotia, and Prince Edward Island, making up the vast majority of Canada's population.

I started this chapter by outlining four categories that individuals and families can explore to help make sure that they are part of the solution rather than the problem. However, there's one other significant contribution that you can make, and that's to *get the message out*. By talking to friends, family members, and coworkers, you can make sure that more people become aware of these issues—just like you're doing by reading this book. You can even suggest they buy a copy to bring them up to your speed, so that you can have intelligent discussions on the various topics connected to global warming and the climate crisis, knowing that you have a working knowledge on most of the pertinent issues. (I suppose in the interests of recycling, perhaps you should consider lending them the book instead, although I expect you'll want to keep your own copy handy for future reference.)

If you're passionate about these issues like I am, you may want to take it further than simply trying to make sure that those around you are aware. You might want to think about even greater contributions to solutions. Perhaps you'll consider donating to organizations dedicated to trying to preserve our planet and its climate. Maybe you'll want to offer your time as a volunteer to their cause. It's possible you'll want to try to make an even bigger impact by petitioning your government to make sure they know that people care and that they need to move toward solutions as well. This can be done at all three levels of government: municipal, state or provincial, and federal. By becoming an educated activist, you can make huge strides in helping to solve the problem.

There is no solution without every person doing his or her own share toward making a change. But the total solution to the climate crisis is truly bigger than that; real strides need to be made beyond individuals, families, and businesses. Changes in industry at the multinational corporate level are required. The solution also depends on changes in government policy and indeed the very fabric of society itself, which currently relies on fossil fuels as the backbone of a global economy.

Key Concepts of Chapter 6

Ways to help reduce your carbon footprint include the following:

- Improving efficiency at home by making smarter purchases and proper recycling.
- Purchasing carbon offsets.
- Investing in renewable sources of energy, both directly in your home as well as indirectly through buying greener electricity.

Suggested Reading for Chapter 6

Clinton, Bill. *Giving: How Each of Us Can Change the World*. New York, NY: Knopf, 2007.

David, Laurie. *Stop Global Warming: The Solution Is You! An Activist's Guide*. 2nd ed. Golden, CO: Fulcrum, 2008.

Kneidel, Sally, and Sadie Kneidel. *Going Green: A Wise Consumer's Guide to a Shrinking Planet*. Golden, CO: Fulcrum, 2008.

McDilda, Diane Gow. *365 Ways to Live Green: Your Everyday Guide to Saving the Environment*. Avon, MA: Adams Media, 2008.

Revelle, Roger, and Hans E. Suess. "Carbon Dioxide Exchange between Atmosphere and Ocean and the Question of an Increase of Atmospheric CO_2 During the Past Decades." *Tellus* 9 (1957): 18–27.

Schumacher, E. F. *Small Is Beautiful: Economics as if People Mattered*. New York, NY: Harper Perennial, 1989.

Vasil, Adria. *Ecoholic: Your Guide to the Most Environmentally Friendly Information, Products and Services in Canada*. Toronto, ON: Vintage Canada, 2007.

Chapter 7

Reducing Greenhouse Gas Emissions: The Big Steps Society Can Take

> We are prone to judge success
> by the index of our salaries
> or the size of our automobiles,
> rather than by the quality of
> our service relationship to humanity.
> —Martin Luther King Jr.

Recall in chapter 3 where we considered the features we'd like to have in an ideal energy source:

- Inexpensive
- Energy-dense
- Easily stored
- Easily transported
- Endless supply
- Clean

Currently, our world depends on fossil fuels as its main source of energy, and fossil fuels fulfill most of the criteria listed here but fail on the last two. It's a known fact that our fossil fuels will eventually run out. The first signs that we have a finite amount are already evident: we're looking to offshore drilling as a civilization now more than ever before. (The BP oil rig explosion and the massive oil leak associated with it in the spring of 2010 are a testament to some of the difficulties we face when we have to go so far down below the ocean's surface simply to reach ground and then have to dig down even deeper from there.) We also see it in the efforts Canada has put into developing the oil sands in Alberta. Although it takes much more energy and cost to obtain a barrel of oil out of these more extreme sources than it does to drill for it in the Middle East—and with significant environmental impact too—it's still a profitable endeavour given that fossil fuels are so interconnected with the global economy.

Even if fossil fuels didn't contribute to climate concerns, our planet would still eventually need to look to alternate sources of energy, perhaps in fifty years, perhaps in five hundred. Thus, the world of the future will use energy much differently than we do today. Given the fact that fossil fuels also contribute to climate change and global warming through greenhouse gas emissions, most who are concerned (such as myself) believe that we should start moving toward these alternate sources of energy sooner rather than later. It doesn't make sense to wait until we are in a crisis situation, whether that crisis is because we're desperately running out of our fossil fuels, or because their use is destroying our climate. Since such a transition is going to take some time, we need to ensure that we are consistently moving in the right direction toward alternate energy sources and away from fossil fuels.

The last chapter listed a variety of things that you and your family can do to minimize personal carbon footprints and help contribute to the solution. Unfortunately, that won't be enough to prevent a crisis in years to come. We have to make changes as cities, as states and provinces, and as nations. Indeed, we have to tackle this on a global scale. No one nation can solve this problem. As the largest contributors to greenhouse gas emissions, the Unites States and China certainly need to make major changes, but every nation must play a part in this solution. Ideally, those two nations will help lead the way once they see fit to make some much-needed changes.

Imagine some day in the future when fossil fuels are no longer used as major sources of energy. How will that future look? A number of changes will need to occur to effectively deal with alternate energy sources because our current infrastructure is designed for a system that uses fossil fuels and not for the sources we'll be using in the years to come.

Energy Use in the Future

First, let's look at the way we use energy today. Currently, the world obtains fossil fuels at remote locations. Those fossil fuels are then sent to facilities where they're refined and made ready for practical use. Ultimately, they are transported to their final destination where they're needed. The gasoline in your car may well have come from the Middle East, been shipped to a refinery in North America, and then been transported to your local gas station where you pumped it into your car; your car then extracted energy by the process of combustion within its motor. The electricity in your home may have come from coal that was mined. In the United States, a number of states are major sources of coal mining, with Wyoming and Kentucky being the two largest producers; in Canada, the provinces of Alberta and British Columbia contribute most. Once coal is mined, a train usually transports it to a power plant, where its combustion is used to create the steam that can rotate turbines and produce electricity. (Your particular home may get its electricity from another source, such as hydro, but most of the world's electricity is produced from coal.)

Power stations along the grid help to regulate the amount of electricity flowing because it's still difficult to store electricity in massive amounts. Most of the time, the amount generated depends on predictions of how much will be needed according to previous usage patterns over the years. Predictions are also in part based on weather forecasts since heating and air conditioning account for the bulk of major fluctuations in electricity use. There's always a little reserve, but it isn't a lot. You've probably experienced a situation, say in peak summer with particularly hot days ahead, where the message from the power companies is heard loud and clear: avoid too much use of your air conditioning and run your appliances at night to avoid excessive electricity usage during peak daytime hours. This is an effort to smooth out the hourly electricity requirements and avoid a shortage.

The energy of the future will be quite different. Instead of fossil fuels that are obtained and then transported great distances, with the energy produced closer to the source where it's needed, energy of the future will be produced where it's acquired because the "fuel" used won't be transportable. The best prospects will be the renewable energy sources: solar, wind, and geothermal energy from deep in the Earth's interior. Since we can't move the sunlight, the wind, and the heat deep within the Earth, the electricity generation from those sources will take place exactly where the renewables are located. As a result, there will still be significant distances to transport the energy to where it's needed, but these distances will be even more vast than they are today because the power stations generating today's electricity are still relatively close to where the electricity is being used.

To effectively deal with this, we will need to have in place a new method of storing and transferring electricity; this is usually referred to as a *super grid*. The idea has been tossed around for decades, but since the need for such a grid hasn't been so pressing to date, the concept hasn't moved forward enough to be ready for our future needs just yet. Some modifications can be made to existing power transmission systems, but for the most part, the systems will need to be replaced. The grid of the future will need to have the ability to transfer much higher voltages much greater distances.

As you can imagine, this transition will have a significant cost attached to it. Also, most people will likely have a "not in my backyard" attitude, so that although they may support the idea of a super grid as part of a move toward greener energy sources, they won't want the storage facilities and power lines to be anywhere near the communities where they live. Health concerns about living in close proximity to such facilities may actually be valid; since such a system doesn't presently exist, we can't properly test its effects on the health of nearby communities. Thus, mapping out the lands where such a grid can be built will come with its own share of hurdles, not to mention the local environmental concerns that will go along with building it.

Another aspect of the electrical grid of the future will be the ability to transfer the electricity in the most efficient and practical manner, often called a *smart grid*. A good analogy is the way in which we watch cable television today. Many years ago, people watched television via the signals that were broadcast through the air, similar to radio. That's why broadcasters were considered "on the air." Those television signals were broadcast using radio waves. Much of what we listen to on the radio is still broadcast this way, although satellite radio is gradually becoming more popular. Also recall that those radio waves are a part of the electromagnetic spectrum. When you enjoyed television in this manner years ago, you watched whatever was on at the time it was broadcast, often using rabbit ears to enhance the signal (if you're old enough to remember them, as I admittedly am). Even cable television for most of its life has been a one-way street; information came into your television, and you watched whatever channels you were interested in. If you didn't find anything you liked, you were out of luck.

The cable television of today is vastly different; it's truly a two-way digital technology. Although the cable transmission still comes into your television, and you can flip through channels just as you did in previous years, now your cable box communicates back to your provider so that you can actually tailor what you want to watch and when. On-demand television is a major benefit from this, so that if nothing currently being broadcast is to your liking, you can watch something of your own choosing. You can select what you want when you want, and your cable box provides the appropriate signals to your television precisely at that time. When you're watching a movie or television program in this way, you may be the only cable subscriber out there who's getting the signal for that particular show at that precise moment. If you need a bathroom break, you simply pause your show, and away you go. You can even see your cable bill on your television screen if you want. All of this is possible because of two-way communication technology.

The smart grid will be conceptually similar to this. Along with the changes required to handle the transportation of electricity within the super grid described earlier, there also will be more advanced digital metering of electricity use at each individual home and business. There will be a differential in electricity cost so that it will be less expensive in off-peak hours, encouraging people to use more electricity in the evenings rather than during the daytime. Also, if homes and families of the future are in the process of generating their own electricity by some of the methods described in chapter 6, then any surplus they don't need can be supplied back to the grid by means of the two-way transfer of power that will be available. The cost of that electricity would then be refunded to the customers who supplied it.

Storage of such massive amounts of electricity will have its difficulties to be sure, but one idea that is getting serious consideration has to do with electric vehicles. If every car on the road is electric, and there are outlets for them at homes and in the parking lots of buildings where people work, then there will be a huge amount of battery storage potential out there connected to the grid most of the time. Regulating the precise back-and-forth transfer of electricity will require significant computing ability, but this is the type of infrastructure that will be required to wean our society away from the use of fossil fuels.

Alternatives to Fossil Fuels

As I stated earlier, the best places to look for alternatives to fossil fuels are the renewable sources: solar, wind, and geothermal. Since two of these are dependent on energy from the sun, and one from the heat buried deep inside Earth, they are for all practical purposes limitless and thus truly renewable. There are other sources we can look to as well, including nuclear, which harnesses the energy of the atom, and biofuels in which we grow a source of fuel. I'll review those a little later, but for now let's focus on the renewables.

Solar

When most of us think of solar energy, solar panels like those I described in the last chapter come to mind. These are the photovoltaic panels. They work because the light energy from the sun obtained from photons hitting these panels induces electrons to start to move, creating an electric current in the process. They don't need to heat up water and convert it into steam to rotate a turbine the way fossil fuels do; the panels get electrons moving and generate a direct current right on the spot. That direct current can be processed through an inverter and changed into an alternating current that can be transported greater distances for use as electricity in the grid. Although we can put these panels on the roofs of our homes or on the tops of commercial buildings, we can also make a large enough array of them to function as a solar farm.

The Olmedilla Photovoltaic Park in Spain is the world's largest such solar farm. It was constructed in 2008 and consists of more than 160,000 solar panels generating enough energy to provide the electricity needs for over forty thousand homes.

An array of solar panels "farming" the sun's energy.

There is another way to use solar energy that isn't as well known as photovoltaic panels. It's called *concentrated solar thermal* (CST) and uses reflected sunlight by way of curved or parabolic mirrors. These mirrors concentrate the sunlight onto a source of liquid, such as water in a tower, which is then heated enough to boil the water and turn it into steam. The steam can then be used to rotate a turbine and generate electricity. Instead of using fossil fuels such as coal to heat water and generate steam, producing greenhouse gases in the process, CST uses sunlight with no emissions produced whatsoever. The PS10—or Planta Solar 10—power tower near Seville, Spain, uses 624 mirrors to direct sunlight to the top of a tower that contains the reservoir of water. It generates 11 megawatts (or 11 million watts), which is enough to supply the energy needs for thousands of homes.

Obviously, solar energy requires sunlight, which means that electricity is generated only during daytime hours. Fortunately, that's when our electricity use is generally at its peak. More significantly, however, it also means that for them to be at all productive, solar power plants have to be in locations that get a lot of sunlight. Both the Olmedilla Photovoltaic Park and the PS10 Power Tower are in Spain because that region of the world gets a significant and reliable amount of sunlight exposure. Getting the electricity to cities in the future will require a super grid like the one described earlier because most people in the world live great distances away from such consistent sources of sunlight.

Mirrors reflecting sunlight to the top of a concentrated solar thermal tower located in Spain.

Wind

Wind is generated by the movement of air masses as they heat and cool, so this is another source of energy available to us that comes from the sun, albeit indirectly. Using wind power on a large scale to generate electricity requires the use of numerous wind turbines. These wind turbines are much different from the windmills of old, but because they're slowly popping up all over the planet, there's a good chance you've seen one. Their method of electricity generation should be apparent; as the wind makes the blades rotate, rotational energy is obtained that can turn the turbine needed to generate electricity. There isn't an intermediate step where water is heated to generate steam as occurs with the use of fossil fuels or CST.

Naturally, wind power will need to be obtained in areas of the planet where good winds exist, similar to how solar power needs to be obtained in regions that get a good amount of sunlight. It can be farmed either from regions of the world where there are stronger winds that vary or from places where there are winds that are less strong but consistent. Once again, that means the electricity needs to be generated in areas that are remote from the cities that will use it. A good example is the Danish wind farm offshore from Copenhagen. In North America, the strongest wind corridor is centrally located from Texas in the south up to North Dakota and into the Canadian prairies. Winds are always particularly strong offshore as well—the very reason the Danish wind farm is located where it is—so the coastal regions in both the Atlantic and Pacific Oceans and the Great Lakes provide potential sources of significant wind power too.

A wind farm located in Whitewater, California.

Once again, transporting the electricity that's generated through a super grid will be necessary for this to be at all practical on a large scale. (Although concerns about injuries to birds flying into these turbines are sometimes raised, the statistics show that it's unlikely that turbines will ever kill as many birds as household cats do. Unless we're also prepared to eliminate cats from the planet as an environmental threat to the avian population, I'm not sure that particular argument against wind turbines is a very good one.)

Geothermal

In the last chapter we reviewed the process of geoexchange that you can consider for your homes, taking advantage of a constant source of heat below the frost line in the ground. True geothermal energy is different and takes advantage of the heat deep within Earth's mantle, coming from some of the latent heat that still exists as a result of Earth's formation nearly five billion years ago, along with some of the heat generated from the radioactive decay of minerals deep underground. It's one of two major sources of energy we can access that do not come from the energy of the sun, the other being nuclear power. Harnessing this source of energy is cost-effective, reliable, sustainable, and generally more environmentally friendly than drilling or mining for fossil fuels. However, it's limited to sources on the planet where we don't have to dig to prohibitively deep levels. These areas are often near where tectonic plates meet. An example of this is a region referred to as the "Ring of Fire," which encircles the entire Pacific Ocean. This same "Ring of Fire" is also responsible for many of the earthquakes and volcanoes that affect our planet, as California and Japan have repeatedly experienced.

The most conventional type of geothermal energy involves drilling into sources of heated water deep within the ground, so-called hydrothermal plants. However, since many sites have the heat but not the water, a modification to the basic concept involves pumping water down to the heated rocks where the water itself becomes heated and is then pumped back up, either by causing some fractures in the rocks themselves or because the rocks are permeable enough to allow the water to flow through them. This is known as enhanced geothermal energy and is currently the focus of a lot of research. The largest geothermal power plant on Earth consists of twenty-two geysers located in Northern California. Unlike the geoexchange process used to heat and cool your home—where the digging needs to go only about two metres or six feet deep—geothermal power plants need to dig several kilometres or miles below the Earth's surface. When the hot water is returned to the plant, it's converted to steam and used to drive a turbine that again generates electricity.

Some of the drawbacks to geothermal energy include the high costs of drilling to such great depths as well as the risks involved in drilling of this magnitude. Also, there isn't always a guarantee that the heat will be exactly where it might be predicted. As with other renewable sources of energy on an industrial scale, it's yet another example where electricity is being produced in one location but needs to be transported to remote regions to be at all practical. For any of these sources of energy to have a possibility of becoming a reality, there must be a super grid in place to accommodate the transfer of massive amounts of electricity great distances.

Other Sources of Energy

For various reasons—not the least of which is that they satisfy the requirements of an ideal energy source better than anything else available—the renewables previously described are the most attractive solutions to an alternative to fossil fuels. However, they aren't necessarily the complete solution to this problem, and other sources may need to play a role, either as part of the transition to eliminating our dependence on fossil fuels or as part of the long-term solution in a sustained manner. These deserve some attention as well.

Nuclear

Nuclear power is already a major player in the production of the world's energy. The United States produces the most, but because it's such a populous nation, with more than 300 million people, nuclear accounts for only about 19 percent of its electricity production, with the burning of coal still producing the vast majority. In contrast, France is quite a bit smaller, so it doesn't produce as much nuclear power as the United States overall, but what it does produce accounts for about 80 percent of the electricity needs of the country.

Today's nuclear power comes primarily from the process of fission. You will recall from chapter 1 how the sun's energy is created from the fusion of two hydrogen atoms into one helium atom, with some energy released in the process, according to Einstein's famous equation $E=mc^2$. Fission is a different process, whereby one atom splits into two different atoms. However, fission isn't simply the reverse of fusion. Rather, they are different types of nuclear reactions, but both release energy. In the case of fission, the energy comes from the breaking and reforming of nuclear bonds inside the atom, along with some kinetic energy as the newly formed materials fly apart with great speed. You'll remember from chapter 4 that with chemical reactions such as the combustion of methane, the energy from the newly formed chemical bonds in the water and carbon dioxide molecules was greater than the energy it took to break apart the original bonds in methane. Nuclear power is similar because the energy derived from forming the new nuclear bonds is also greater than the energy required to break apart the initial bonds, but nuclear bonds always contain much more energy than chemical bonds do.

Although nuclear energy avoids the problem of greenhouse gas emissions, uranium mining and nuclear waste have their own environmental concerns.

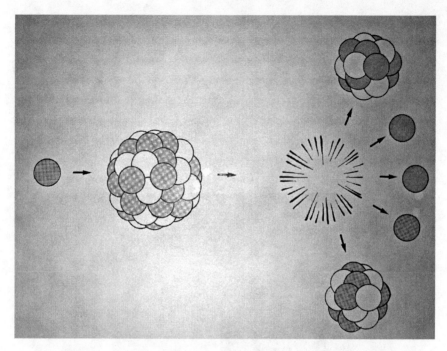

An example of nuclear fission: a neutron collides with a uranium atom, producing two smaller atoms and releasing energy in the process; three neutrons are also released, allowing the reaction to continue. (Illustration courtesy of Dr. Rob Ballagh)

An example in common use today is an isotope of uranium known as uranium-235. Recall that isotopes of an element have the same number of protons but different numbers of neutrons in the nucleus, leading to different atomic weights. In this case, the number 235 refers to the total number of protons and neutrons combined, in other words its atomic weight. Uranium's atomic number of 92 tells us that it has 92 protons, so the uranium isotope with an atomic weight of 235 includes 143 neutrons. By comparison, uranium-238 also has 92 protons—otherwise, it wouldn't be uranium—but it has 146 neutrons, for an atomic weight of 238. (As atomic numbers get higher, it's more likely that there will be different isotopes found naturally for the different elements, but as discussed in chapter 5, even elements such as hydrogen and carbon can have different isotopes.) If an atom of uranium-235 is hit by a neutron, which then combines with all of the other protons and neutrons located in its nucleus, the atom will briefly become uranium-236. It will then split into smaller products with atomic numbers that usually add up to close to the 235 but not quite. This is because neutrons are freed up in the process. For example, uranium-236 can divide into krypton-92 and barium-141, with three neutrons released in the process (92 + 141 + 3 = 236). Other fission products can be formed as well, and this is only one example.

The process is rather complicated, and a more detailed explanation is beyond the scope of this book, but there are various types of radioactive decay. (You are already familiar with one example, previously described in chapter 5, known as beta decay. Beta decay is where neutrons are converted into protons, and it's how carbon-14 decays into nitrogen-14, making it useful for carbon dating.) None of the specific types of radioactive decay need to be understood in any detail for our purposes at present. What is important to appreciate, however, is that a neutron colliding with a radioactive atom starts the whole process, leading to the fission of uranium-235 into smaller products and freeing up some neutrons as a result. As long as there's more uranium-235 around, then the neutrons that are freed from the collision can continue the process, leading to a chain reaction. (The plutonium used in nuclear weapons takes advantage of this process as well. However, those who use plutonium in such a manner have no intention of controlling the nuclear reaction that results.) Although it takes some energy to break the nuclear bonds and split the uranium atom, the energy obtained by reforming the nuclear bonds in the products of the fission process more than makes up for it, so that a net gain of energy is obtained. That energy can be used to heat water and turn it into steam that can then rotate a turbine and generate electricity.

What keeps the reaction under control in a *nuclear reactor* is the use of something known as control rods. These are materials that can easily absorb the stray neutrons freed up in the nuclear reaction. Many different materials can be used in control rods; common materials include the elements silver, indium, and cadmium. If the reaction needs to increase, the control rods can be partially removed; if the reaction needs to be reduced, the control rods can be inserted back in. In this way, a chain reaction getting out of control can be prevented. Of course, it isn't always 100 percent foolproof, as the accidents at Three Mile Island in Pennsylvania in 1979 and Chernobyl in the Ukraine in 1986 demonstrated. Those particular disasters were largely due to human error, but the tsunami that affected Japan in the spring of 2011 proved that natural disasters can also cause significant problems; in that event, at least three nuclear power plants in Japan exploded because of cooling system failures resulting from the structural damage caused by the earthquake and subsequent tsunami.

Nuclear energy also has other significant drawbacks. One is the environmental harm in mining for uranium since mining always causes significant damage to land and the habitats that depend on it. Another is the production of radioactive waste as a by-product. Although a by-product such as cesium-137 can find a purpose in medicine—it can decay to barium-137, a material used in hospital radiology departments for imaging tests—much of the degradation leads to materials that are still radioactive and will continue to decay but have no practical use either in an industry such as health care or as a nuclear fuel. Even though nuclear energy doesn't produce greenhouse gas emissions, proper disposal of nuclear waste will always be a problem for society to contend with. Many consider the use of nuclear power to be nothing more than swapping one environmental concern for another.

Biofuels

Biofuels are derived from biomass or organic matter. Although fossil fuels ultimately came from organic matter as described in chapter 2, they don't count as biofuels because of the processes of extreme heat and pressure they were subjected to, which produced the fuels only after many millions of years. Biofuels come from plant matter that is grown via more conventional farming processes, but in this case, the materials grown can be used as fuels, such as vegetable oils, alcohol, and even biogas. Since the growth of new plants to replace the ones already used for fuel takes only a few years rather than the millions of years required for fossil fuels, many consider biofuels to be a renewable resource. They ultimately derive energy from the sun just like solar and wind energy, only this time the energy is obtained indirectly by way of photosynthesis instead.

People have been using biofuels for thousands of years already: woodburning has been a major source of energy for heating and cooking since the Stone Age. Although burning wood generates greenhouse gas emissions, trees can be grown to replace those that have been cut down and used as a source of fuel. In a sense, the carbon dioxide generated by burning wood can be reabsorbed in just a matter of years through photosynthesis if those trees are replaced with newly planted ones. Ultimately then, there doesn't have to be a net increase in global carbon dioxide. Peat has also been used as a biofuel. (You will recall that peat is what can ultimately lead to coal if the conditions are just right and you wait long enough.)

Biofuels can be solid, liquid, or gaseous. Examples of solid biofuels include wood, charcoal, peat, and manure. Liquid biofuels include alcohols such as ethanol and biodiesel. The main biogas is methane, which is generated from the anaerobic digestion of organic matter. Methane will act as a greenhouse gas if it's accidentally released into the atmosphere, however. (Recall that bacteria in the stomachs of ruminants such as cattle and sheep also generate plenty of methane that is released eventually, from one end of the animal or the other.)

Ethanol gets significant attention as a potential replacement for liquid fossil fuels, such as gasoline, because the combustion of ethanol leads to less greenhouse gas emissions than gasoline does. This is because ethanol has only two carbon atoms in its chain, whereas the octane in gasoline contains eight, so ethanol is closer to the cleanest fossil fuel, which is methane, producing relatively more water compared to carbon dioxide from its combustion.

Ethanol is already in use today as an additive to gasoline. The next time you fill up at the pump, you will probably be able to spot a sticker somewhere that points out that up to 10 percent of the fuel you are purchasing may contain ethanol. Ethanol has had its share of controversy, however, because the first generation of ethanol production as a biofuel was based on corn, which requires the combustion of fossil fuels in its production such that net emissions were really no better than if we simply used gasoline in the first place. Second-generation ethanol production will use other materials such as switchgrass—a tall grass that grows throughout much of North America—and it is anticipated that this will result in even lower emissions, but that technology is still in development. Brazil is one country that uses large amounts of ethanol as a biofuel; there it is generated from wheat and sugarcane, along with some other crops.

Another potential fuel is something referred to as algal fuels. Algae are the main plant life in the oceans. Because our oceans are so vast, covering about 75 percent of our planet, algae—and not the plants found on land—are responsible for most of the photosynthesis on Earth. Currently, however, there are efforts to use algae as a source of fuel by growing it on land too. Early studies suggest algae may be able to produce about thirty times more energy per acre than other biofuels. Algal fuels also would have the advantage of low toxicity with little threat to the environment if there were ever a spill since it's nothing more than plant material. Like most biofuels, this option is still in development and is not yet ready for primetime commercially.

For biofuels to be a practical source of energy in our future, a number of factors will need to be addressed. Obviously, there shouldn't be substantial amounts of carbon dioxide generated in the process, and the growth of biomass should have a low risk for harming the environment. Also, it's better if the biomass used for fuel is different from crops that are grown for food. Otherwise, there will be a risk of increased food costs resulting from the competition that would arise.

Carbon Capture and Sequestration

As you can appreciate, fossil fuels aren't likely to disappear anytime soon. Truthfully, it's unlikely that anyone alive at the time of this book's publication will live to see a world where fossil fuels no longer play a significant part. That will occur one day since fossil fuels will inevitably run out, but nobody is sure exactly when that will be. Despite their drawbacks, fossil fuels play a major part in the global economy, and the world's energy infrastructure depends on them. The wheels of progress tend to turn slowly, especially when money and politics play a part. Also, some of the nations with the fastest-growing economies, such as China and India, are very dependent on them at present. Likewise, the United States, which also uses large amounts of fossil fuels, is unlikely to abandon them too quickly when the changes needed to switch to some of the alternate sources of energy described here require money, effort, and agreement across party lines.

Since fossil fuels will be around for a while yet, it makes sense to develop ways to make them cleaner, if possible. The method that has been most researched is carbon capture and storage or sequestration. The basic goal here is to take the carbon dioxide out of the picture before it has a chance to be released into the atmosphere and store it somewhere safely. The usual place considered for such storage is deep underground, even beneath the ocean floor.

The first required step is to remove carbon dioxide from the gases released in the smokestacks, a process known as scrubbing. That technology has been present for many decades. (For example, the carbon dioxide exhaled by the Apollo astronauts was scrubbed to avoid dangerous levels building up inside their capsules.) Once removed, it needs to be transported, most likely via pipelines, to the region where it will be stored or sequestered. At its final destination, the carbon dioxide is compressed into a liquid state and then pumped deep into the ground.

This can be accomplished in a variety of ways. For example, it can be pumped into regions where oil already exists and is being drilled for, with the carbon dioxide that is pumped deep into the ground helping to displace the oil, making it easier to bring the oil up to the surface. This process is known as enhanced oil recovery. Another option is to pump it into porous rock, with the goal that it will remain there, hopefully for millions of years. In other words, we would be helping to return carbon back to where it came from originally—before we got our hands on it—aiding Mother Nature's carbon cycle along.

As you can likely imagine, there are a number of concerns with the whole idea of carbon capture and storage, making it unlikely that the process will be a major contributor to a solution anytime soon. And because such a process costs money, it's unlikely that industry will make much effort to effect such changes unless mandated by government, which in turn is not likely to occur until the world's governments all come together and place a price on carbon. Simply put, many industries today do what they do, spewing tons of carbon dioxide into the atmosphere, because there's no need for them to give consideration to their emissions. If, however, there was a price on those emissions so that businesses that found ways to reduce their emissions could keep more of their profits, whereas those that continued to do business as usual would be hit with a penalty or tax for their higher emissions, then industries would soon find the means to reduce their emissions as best as possible in the name of improving profits.

This idea has found its way into most of the international treaty discussions that aim to tackle the climate crisis. It might be as simple as this: if you release a lot of carbon dioxide, you pay a fine; if you capture and store it, you get a rebate. One criticism raised is that this wouldn't really solve the problem because less energy would be produced from, say, coal-burning power plants since more energy would be put into the capture and storage of the carbon dioxide. Since the electricity needs of cities would be the same, these power plants would have to burn even more coal to equal the amount of energy produced when carbon capture and storage wasn't enforced. And of course, there will always be the environmental concerns about storing highly pressurized liquid carbon dioxide deep underground and beneath our oceans and the possible risks associated with that. However, this is still considered safer than the spills associated with petroleum products for the simple reason that presently, without carbon capture and storage, all of that carbon dioxide is making its way into the atmosphere anyway. This technology, like the others described previously, requires more research and development before we can consider it a useful part of the global solution.

Governmental Policies

Governments have a number of opportunities to make a real difference in greenhouse gas emissions and ultimately make a positive impact on climate change. Governments can consider incentives to encourage the population to make changes voluntarily, like some of those described in chapter 6 (e.g., installing solar panels and geoexchange systems and purchasing hybrid or electric vehicles). Governments can also implement the sorts of projects discussed in this chapter and put a price on carbon that would discourage the use of fossil fuels in industry.

Incentives are easy: all a government has to do is consider some financial break for those individuals, families, and businesses that choose to reduce their greenhouse gas emissions. As already pointed out, such changes usually cost money. Those who makes those changes without incentives tend to be people and businesses who (a) have the financial means, (b) believe that the changes are money well spent, and (c) are motivated to do what they believe is the right thing. If you could make it a more fiscally attractive option, then many more people and businesses would jump on the bandwagon and retrofit their homes, consider solar or geoexchange energy, and purchase hybrids—and ultimately electric vehicles once they are more readily available.

There are many out there who have the motivation but not the financial means to comfortably make these changes. Governments can assist in that regard with tax breaks and rebates. If people have to spend another $20,000 on a vehicle simply to choose the greener hybrid option, few will consider it. But if the government provided a rebate for buying such a car, then more would consider the option. Obviously, the bigger the rebate, the more people who would think about it. Slowly, these sorts of policies are coming into play, but if governments are going to be serious about encouraging their citizens to voluntarily make a difference, they need to help with financial incentives.

Taking the technologies described in this chapter and putting them into practice is a slow process. Often, there are lobbyists—industries dealing in fossil fuels are among the most powerful—who have plenty of pull when it comes to making, or preventing, policy changes. If the oil industry has consistently provided support for certain politicians or a particular political party, then there is always a threat, whether real or perceived, that such support will be withheld if the government makes any change that might hurt the very industry that helped get those politicians elected in the first place. This is the reality of human nature, and unfortunately, the power of the almighty dollar has created far too many impediments to progress over the centuries for modern civilization. But if governments pursue the proper research and development so that they can seriously implement any of the green forms of energy available on a large scale, as well as the super/smart grid necessary to deal with that energy, then significant differences can be achieved.

We face an interesting dilemma if we hope to convert to cleaner sources of energy, and you may have already thought of it. Imagine that every nation in the world has come together and agreed that we need to make changes and convert to clean energy as soon as possible. Thus, wind farms are built with rows upon rows of wind turbines; solar panels and concentrated solar thermal farms are constructed; and geothermal energy sources are tapped. Great efforts are put into place to build the super/smart grid necessary to cope with these new sources of energy. Have you spotted the problem yet? In order to get from here to there, we have to invest a significant amount of energy in the manufacture of everything required to make the world of the future a reality. However, we have to do it with the energy sources of today. In other words, building wind turbines, solar panels, geothermal power stations, and newer power grids during this transition process will produce a significant amount of greenhouse gas emissions, and there's no getting around that fact.

Of course, once the new system is in place, any future manufacturing will be done with clean energy, so it won't be a perpetual problem. But as it stands, it will take some time for a wind turbine of the present to pay itself off in saved greenhouse gas emissions, given the amount of carbon dioxide produced to manufacture it in the first place. Those who oppose moving toward greener sources of energy often use arguments such as this, but if we continue with the status quo, then we're continuing to produce emissions with no end in sight. If we at least pursue the transition, we will indeed add to the emissions, but once we've achieved our goals, that problem will diminish. The transition will contribute, but the final result will be worth it—short-term pain for long-term gain.

Global Efforts at Solving Climate Issues

International treaties have been developed over the years to try to effect beneficial changes regarding environmental concerns, such as pollution. Some have been particularly successful. One such treaty was between the United States and Canada and was signed in 1986 in an effort to reduce acid rain. The Acid Rain Treaty was largely successful because it involved only two countries with a longstanding history of amity and trade, and it tackled only one component of pollution.

Another successful treaty was the *Montreal Protocol*; it involved countries from all over the world and was enacted in 1989. Its purpose was to reduce the use of chlorofluorocarbons (CFCs), which have been used commonly as refrigerants, propellants in aerosol cans, and solvents. They are known to damage the ozone layer, a component of the atmosphere that helps protect us from the harmful effects of ultraviolet radiation. This effort was successful because it targeted only one particular group of pollutants even though many countries were involved. Although an important part of industry for most of the twentieth century, CFCs were relatively simple to phase out because the economic framework of our society could still survive largely unscathed without them. Phasing out CFCs involved billions of dollars, but phasing out fossil fuels will cost trillions of dollars. Because of the significant role they play in the global economy, the situation with fossil fuels is therefore going to be much more difficult to tackle than either acid rain between two friendly nations or CFCs among many nations. But efforts have been made.

In 1992, the first international effort to address climate change occurred at the World Summit, with attendance from most world leaders. Out of that summit came an important document known as the United Nations *Framework Convention on Climate Change* (FCCC). It acknowledged that global warming was a serious issue that needed worldwide attention and established an Intergovernmental Body on Climate Change to follow the issue, involving many of the world's scientists.

The FCCC recognized that there are differences in economic strength between developed and developing nations. It decided that nations that were still in earlier phases of development shouldn't be expected to invest a portion of their economy to tackle the problem of climate change. The developed nations, however, were considered fair game; they were designated as Annex 1 nations and included the United States, Canada, all of Europe, Japan, Australia, and countries from the former Soviet Union. The Annex 1 nations were expected to take the lead in reducing greenhouse gas emissions, and goals were set to ensure that emissions in 2000 didn't exceed those from 1990. The FCCC was ultimately ratified as international law in 1993 and was unanimously passed in the US Senate in 1997, suggesting that all nations involved, including the United States, were committed to helping solve the problem.

The *Kyoto Protocol* was the first treaty after the FCCC and was created in 1997 in Kyoto, Japan. It was enforced in 2005 once a minimum required number of nations signed on, committing at least 55 percent of the world's emissions from Annex 1 nations to the protocol. European countries quickly embraced the protocol and committed to reducing their emissions substantially. However, by the time Kyoto came along, the emissions from the United States had already increased 8 percent compared to 1990 levels, and the US government felt it was impossible to ratify the protocol and achieve the required reductions without doing substantial harm to its economy. By 2010, 191 countries had ratified the Kyoto Protocol, but the United States is still not among them.

The countries that have ratified the protocol have different targets for emissions. For example, the European Union has a target of reducing its emissions 8 percent by 2012 compared with 1990 levels; countries from the former Soviet Union have a target of keeping emissions at 1990 levels; and Australia was actually allowed to increase its levels by 8 percent because of how low its emissions were to begin with. Those designated by the FCCC as developing nations still have no obligations to meet, however.

The US government expressed concerns that it was unfair to expect it to make efforts to reduce its own emissions while developing nations, such as China and India, still had no such obligations, even though those nations still contribute substantially to the global problem. This is contradictory, however, to the fact that the United States initially ratified the FCCC, agreeing that developed nations would tackle the problem first.

Canada ratified the Kyoto Protocol with a target of reducing its greenhouse gas emissions 6 percent by 2012 compared with 1990 levels. However, without any effective government policies in place that would have any possibility of making a substantial difference, emissions in Canada have continued to increase with no hope of achieving that target. Also, the Canadian government has been hesitant to make such policies mandatory until the United States does the same. Otherwise, it believes that Canada would be at a significant economic disadvantage because its largest trading partner wouldn't have the same rules in place; manufactured goods would thus be much cheaper south of the border, and the Canadian economy would suffer unfairly. Clearly, until everyone is on board and willing to make changes collectively, no one nation wants to stick its neck out before all others.

The next significant effort at international agreement was the *Copenhagen Accord* drafted in Copenhagen, Denmark, in 2009. Unfortunately, it failed to achieve any significant progress in tackling climate change. It endorsed the continuation of the Kyoto Protocol and contained a number of "motherhood statements" that comment on the seriousness of the problem, often stating the obvious, but agreements were difficult to attain, with a number of nations somewhat distrustful of each other. It seemed to come down to many of the participants feeling that their countries were expected to take on a bigger hit than the others sitting at the same table. As a result, nothing binding came out of it. Nevertheless, countries representing 80 percent of worldwide greenhouse gas emissions have made pledges to reduce their emissions by various levels as a result of their participation in the Copenhagen Accord, and any efforts made are at least a move in the right direction.

These international treaties also give consideration to a cap-and-trade system. This is a simple concept of emissions trading that was first developed in the Kyoto Protocol: nations that sign and ratify the treaty agree that they will keep their emissions below a certain cap or threshold. Not every nation can comply as easily as some others, however, so those that are well below their cap can sell off their unused portion to nations that exceed their own cap. This is the "trade" part of the concept. Thus, if one country exceeds its cap, it can purchase the unused portion from another country that is also a signatory of the treaty. Obviously, there are economic advantages to being below the cap, and those countries that exceed their cap will have a significant incentive to reduce their emissions and save their money. This approach allows the principles of economic theory and a free-market system to encourage the changes desired on a more voluntary basis. Of course, the fact that it's truly voluntary and not binding is its biggest drawback; if a nation chooses not to sign such a treaty, it remains free to do what it wants.

Much of this chapter has been devoted to reviewing technologies that are largely well developed and generally already in place somewhere in the world. However, many technologies are being investigated to help with the problem of greenhouse gas emissions but are currently in more experimental levels of development. If you pick up any magazine on science and technology today, there's a good chance that there will be some reference to these novel approaches being considered, perhaps even feature articles. This certainly demonstrates that society seems to recognize the concerns about continuing on with our present course and that efforts are being made to explore technologies that are viable both economically and politically.

Attitudes are moving in the right direction, but there are some major impediments to making dramatic changes to the way we use energy, as the lack of success through international treaties has demonstrated. Most people who are keen on making changes quickly find themselves frustrated by the slow pace with which society as a whole is embracing the problem. Crucial to addressing this part of the problem is the ability to understand the difficulties we face in trying to change the attitudes of an entire planet.

Key Concepts of Chapter 7

- Renewable sources of energy, such as solar, wind, and geothermal, will play a major part in the energy solutions of the future.

- Other possible contributions to the solution of tackling greenhouse gas emissions include nuclear energy, biofuels, and carbon capture and storage.

- The solution will require a super grid that can handle the transportation needs of future energy sources.

- Nations will need to come together and agree in future treaties to commit to solving the problem on a global scale.

Suggested Reading for Chapter 7

Gore, Al. *Our Choice: A Plan to Solve the Climate Crisis*. New York, NY: Rodale, 2009.

Krupp, Fred, and Miriam Horn. *Earth: The Sequel. The Race to Reinvent Energy and Stop Global Warming*. 2nd ed. New York, NY: Norton, 2008.

Rand, Tom, and Dave Clark. *Kick the Fossil Fuel Habit: 10 Clean Technologies to Save Our World*. Toronto, ON: Eco Ten, 2010.

CHAPTER 8

Progress Is Slow: Understanding
the Resistance to Change

> The ultimate test of man's conscience
> may be his willingness
> to sacrifice something today
> for future generations
> whose words of thanks will not be heard.
>
> —Gaylord Nelson, co-founder of Earth Day

Gaining acceptance for a new way of life is always difficult for the simple reason that it's easier to stick with what you know. This is certainly evident when we make efforts to try to protect the climate. Changing attitudes and policies on global warming is a slow and difficult process.

RESISTANCE TO A HEALTHIER LIFESTYLE

In many ways, I find there are similarities in health care; as a cardiologist, on a daily basis I see people display a resistance to change despite the clear benefits of that change. Admittedly, heart disease and global warming are two completely different beasts, but they share one thing in common: the need to modify attitudes and behaviours that are ingrained in order to avoid outcomes downstream that will prove to be devastating. I'll address how this applies to the resistance of society toward policy changes meant to protect the climate a bit later. I think I can illustrate the point more easily by first addressing the smaller-scale issue of people making decisions about trying to live in healthier ways and the resistance they exhibit to doing so despite the benefits to their health if they make such changes.

As an example, I deal with a number of patients who have atherosclerosis, a generalized narrowing of the arteries caused by plaques that consist of cholesterol, fibrous scar-like tissue, and inflammatory cells. In the coronary arteries of the heart, these plaques are the principle cause of angina—a chest discomfort due to a lack of blood flow to the heart—where a narrowing in an artery prevents the necessary increase in circulation during times of increased cardiac workload, such as physical activity and emotional stress.

These plaques are also the primary cause of heart attacks when one of them is unstable enough to rupture, leading to either a partial or a complete blockage of blood flow to that portion of the heart muscle supplied by the diseased artery. In that case, heart muscle downstream from the blockage loses all of its circulation and dies, permanently lost and replaced with scar tissue, leading to a decrease in the overall pumping action of the heart. If blockages occur in the arteries to the neck or within the brain, this same process can lead to cerebrovascular accidents such as transient ischemic attacks (TIAs), which tend to involve a relatively quick and complete recovery, or strokes with associated damage to the brain that is more likely to lead to permanent disability. If atherosclerosis affects the arteries in the legs, it can cause significant discomfort when patients walk because the leg muscles don't get the amount of blood flow they need, a symptom known as claudication.

Atherosclerosis and cardiovascular disease contribute significantly to death and disability all over the world, generally rivaling all cancers combined for the distinction of being the number one cause of mortality in developed nations. There are a number of risk factors that contribute to its development. Some, such as age, gender, and genetics, are beyond our control. Other risk factors are conditions that we can acquire but that with medications and healthy living we can modify; these include high blood pressure, diabetes, and elevated levels of cholesterol. And then some are predominantly related to lifestyle choices that, with proper effort and motivation, can truly be eliminated, such as smoking, lack of exercise, unhealthy diet, and excess weight.

Many of my patients have had devastating cardiac events, such as heart attacks or even a cardiac arrest, where they needed to be revived with cardiopulmonary resuscitation (CPR) and electrical shocks to the heart to get it going again, known as defibrillation. Those who survive often go on to have an angioplasty to balloon open the diseased artery and have little wire meshes called stents placed within those arteries to help keep them open. Some need bypass surgery to provide blood flow past the blockages. Some even have a pacemaker or implantable defibrillator inserted into their chests to help keep them going and live another day.

And yet, many of these same patients who have been given a second chance at life continue to practice the same unhealthy lifestyle choices that got them into trouble in the first place. Although a few manage to change their habits for a short time, most patients don't change permanently. They may be motivated to make some different choices to get healthier, but usually, they don't stick to them long-term. Some of my patients have not only needed bypass surgery for their heart condition but have also had part of a lung removed because they had lung cancer too; in other words, they've had two different diseases with direct links to their smoking, and yet they still can't give up the habit. I agree that smoking has a component of addiction that complicates the process of quitting, but that's only part of it. Getting into an exercise program, losing weight, and eating healthier all lack the same long-term success as smoking cessation, so it's not all about addiction.

It always amazes me to see this, but because sticking with bad habits is more the rule than the exception for these patients, there must be something to explain why they can't make the changes necessary to reduce their chances of having similar problems in the future. This aspect of trying to change unhealthy behaviours in our patients— "teaching old dogs new tricks"—and being unsuccessful in our efforts is, in my opinion, the most challenging, frustrating, and largely unsuccessful aspect of health care today. If we can't get those who have already suffered the consequences of their unhealthy behaviours to change (what is known as secondary prevention), what hope do we have to get people to make changes to avoid their development of a similar problem in the first place (known as primary prevention)?

In general, most people have a hard time altering well-established routines. Change is tough, and the easier path is always to continue on with business as usual. This is despite the fact that the short-term pain of choosing the tougher path of change is usually a trade-off for the much more beneficial long-term gain. Staying the course has the short-term benefit of getting to avoid the struggles of change but lacks all of the benefits associated with that change downstream.

Again, think of smoking: the short-term gains of *not* quitting consist of avoiding the struggles of withdrawal, holding on to a major source of stress relief, and perhaps maintaining relationships more easily with those friends and family members who are also smokers. The long-term gains of quitting, however, are generally more beneficial overall: these include the cost savings of no longer paying for cigarettes, getting off a substance that is truly an addiction rather than simply a habit, the cleaner air and nicer smell in the house and on clothes and hair, and the significant reduction in risk of cardiovascular disease, cancers, and chronic lung disease.

From a purely intellectual perspective, there really is no argument as to which is the better choice, and yet only a minority of smokers choose to break the nasty habit. Most quitters are unsuccessful in their first attempts because they fall back to their old routines, either because the addiction is too strong and they simply can't give it up or because the next time they're faced with a significant stressor in life, they can't think of a better way to handle it than to cope by smoking cigarettes. (Statistics show that most smokers, on average, aren't successful at quitting until their seventh attempt.)

WE CHOOSE WHAT MAKES US HAPPY

Every one of us is subject to this kind of battle, whether with diet and exercise or with other behaviours, such as being responsible with money. Without realizing it, we usually decide whether we want to do certain things based on whether or not those things make us happy. In the example of smokers, they make the decision to continue to smoke by considering the pleasure derived from it. Smoking makes them feel good—in large part by feeding a nicotine addiction—even though it costs money, causes clothes and hair to smell bad, prematurely ages skin, and is less and less acceptable in public these days. The benefits smokers derive are offset, of course, by the risk of developing a smoking-related disease in the future. Such a decision, often referred to as a utility quotient, is not necessarily rational or even done on a conscious level, but these sorts of factors do play a part in the decisions we make. It is this sort of utility quotient that causes my patients to continue smoking even if they have already required bypass surgery and a lung resection for their lung cancer. Their personal utility quotient calculates that the short-term pain of quitting is worse for them than the long-term gain of being an ex-smoker.

Some famous philosophers, such as John Stuart Mill and Jeremy Bentham, explored the ideas of utilitarianism in great detail. Bentham stated that a utility quotient would take many factors into account to determine what the best course of action is. In calculating how good (or bad) a decision might be, we can take into account the intensity of the outcome (just how good or bad), how long the outcome will last, the probability of experiencing the outcome in question, and the number of people affected by the outcome.

If we look at the example of cigarette smokers, they can use this approach to explore whether or not to quit. Quitting will be tough and often leads to cranky behaviour, with loss of a major source of stress relief, that might last for many months. Perhaps that will affect some of their family and friends as well because they in turn will have to put up with nasty behaviour as these ex-smokers go through the process of quitting.

Compare that with the choice of not quitting and the possible outcome of, say, dying of lung cancer. Obviously, death is a major outcome and a permanent one. However, the probability of getting lung cancer is much less certain. It's by no means guaranteed that smokers will eventually develop lung cancer, but if they do, it will affect not only the smokers who died but also all of those who loved them and will miss them when they're gone. Smokers can place a certain value on each of these factors and decide for themselves whether to quit.

Clearly, most people don't give any conscious consideration to such reasoning or make such a calculation, but it helps to illustrate why some decisions we make may be considered irrational to many and yet still occur. Why do I even address this issue? Because it is precisely these same sorts of factors that we need to consider in understanding why people and governments may seem slow, if not completely at a standstill, when making decisions about possible changes that will help future generations have a better chance of living on a healthy planet.

Resistance to Making Changes to Benefit the Climate

Let's now take the factors that go into the utility quotient and apply them to the issue of whether or not to make changes that will benefit the climate crisis and global warming.

1. *How good is the decision to fight climate change?* Well, it will give future generations the best chance to live on a planet that is comparably as healthy as ours is now. All of the disasters that we fear, such as coastal flooding, more extreme weather patterns, the spread of disease, and the potential loss of life on an unimaginable scale, can be avoided. The exorbitant costs associated with those disasters would also be saved. And we can convert our planet to a global economy that relies on renewable resources rather than one that depends on something with a finite supply, such as our fossil fuels. The decision to continue on with our present state is the easier path, just like continuing to smoke is, but the long-term benefits of that decision are pretty difficult to see.

2. *How long will it last?* The changes required to convert our planet to one dependent on renewable resources will likely take decades, if not perhaps even a century or more. Of course, some people and businesses will benefit economically from these changes, such as those directly connected to the newer sources of energy we will be using. Some, however, won't benefit unless they adapt to these changes. If an oil company is not prepared to provide renewable energy that is greener than fossil fuels, it will be left behind in this transition. In that case, all of the company's employees will eventually be out of work when the company has to declare bankruptcy. This is a part of progress, though, however you define it. There already exist many examples of businesses and industries affected by our twenty-first-century living. Print journalism, for example, has had to adapt to an Internet-connected world. The music industry has had to contend with digital and instantly downloadable music, along with the file-sharing piracy that goes with it. Of course, some industries simply can't adapt and are forced to become defunct and extinct. I imagine

a representative for slave traders lobbying the American Congress in the mid-nineteenth century prior to the Emancipation Proclamation: "The livelihood of me and my family and the similar plight of every other slave trader would be compromised if slavery were to become illegal. I'm doing something that my father and grandfather have done before me. The success of the United States was due in large part to this industry. How could the government seriously consider turning its back on us now?" History tends to be cruel to those who are unable to adapt to those changes that are clearly in the best interests of society as a whole, and not simply the select few who are really in it only for themselves but who choose to hide behind such spurious arguments. In other words, society tends to do the right thing eventually. With the devastating consequences we'll face as a result of our greenhouse gas emissions, we have to make sure that we act soon enough to avoid disaster.

3. *How certain is it?* Well, here's where a lot people will disagree. Many try to argue that global warming and the climate crisis are a myth; to the skeptics, all I can suggest is that they reread the chapters in section 2. However, it's fair to say that there is some uncertainty as to exactly when our planet will be in a crisis if we continue down the same path. Most adults today seem to have noticed differences in the summers and winters compared to their childhood, and we hear about weather records being broken regularly, but it is prudent to be careful before jumping to the automatic conclusion that these are definite signs of the climate crisis. (This is exactly what the skeptics do every time there's a massive snowstorm, channeling Edward G. Robinson in *The Ten Commandments* and asking loudly, "Where's your global warming now?" The irony there, as I outlined in chapter 5, is that more severe snowstorms are predicted by the science behind global warming. If these skeptics will not be convinced until we reach the point of no more snowfall, it definitely will be too late to fix this problem.)

4. *How many are affected?* This one's easy: everyone on the planet. If we choose to make changes toward renewable sources of energy, everyone living will be part of that transition. There will be some

hardships, just as there are for the smokers who are struggling with quitting. Some will be hurt by the transitions in industry as described in point 2, and some will benefit. The largest benefactor will be future generations because they won't be left with a world that is far less friendly or habitable than the one we inherited from the generations who came before us. If we continue on our current path, however, then people will see the harmful consequences of that decision, but it will be our grandchildren and their grandchildren who will truly suffer from the poor choices we have made. I expect they'll wonder why we didn't fix the problem while we still had the chance.

Sustainable Development

The term "sustainable development" refers to the concept of balancing our energy requirements with those of future generations. In 1987, the World Commission on Environment and Development defined sustainable development as that which "meets the needs of the present without compromising the ability of future generations to meet their own needs." This comes down to the very heart and soul of what polarizes individuals, industries, and government on whether to make changes. Those who favour change toward renewable sources of energy argue that if we don't make these changes, we will leave to our descendants a planet that has been irreparably damaged by the folly of our excesses and our selfish interests. Those who favour the "business as usual" approach argue that if we do make those changes, we'll hurt the present economy, and the people of today will suffer for the sake of the people of tomorrow, perhaps unnecessarily because they don't consider the levels of certainty to be high enough to justify such drastic changes.

Although significant sacrifices have been made in the past in the name of future generations, it has generally been in the context of something as drastic as war, removing a very real and present immediate threat that is easily identified. Those we call "the Greatest Generation," who fought in World War II, with millions losing their lives, will never be forgotten because of the better world they helped to shape. But are we truly facing a situation like that with the climate crisis, such that we'll possibly sacrifice our global economy, making today's world harder to thrive in, for the sake of the world of tomorrow? *Sustainable development* looks to find the right balance so that those living now and those living in the future can thrive equally without any one group unfairly or excessively hurt by the decisions made.

Sustainable development looks to both economic and social aspects of the decisions and policies that are made by governments, with the environment playing a key part in the whole equation. Take a look at Canada's Athabasca oil sands near Fort McMurray, Alberta. Also known as tar sands (although they don't contain any tar but rather the tar-like substance known as bitumen), they are the largest reservoir of bitumen on the planet, containing about 13 percent of the world's proven oil reserves. Although commercial projects in Athabasca have been in place since 1967, the technology has allowed for a significant acquisition of oil only in recent years. That's because it isn't like oil that comes up from a gusher in the Middle East, ready for the refinery. Rather, it has to be mined—just like in most other open-pit mines—and then crushed, with hot water added; separated into its different components; and cleaned so that the sand can be removed.

The Athabasca Tar Sands in Alberta, Canada, contain large amounts of bitumen that can be converted into oil.

In other words, much more energy has to be invested—along with greater amounts of greenhouse gas emissions generated—to get a barrel of oil from the Athabasca oil sands than to get a barrel of oil from the Middle East. A significant amount of the natural environment is damaged in the process as well, just like with any mining operation. As you can well imagine, many environmentalists are vehemently opposed to these operations. These environmentalists aren't only from Canada because it's in their own backyard but are from all over the world. They argue that the oil sands produce the dirtiest oil around, with the largest carbon footprint of any petroleum product on Earth, given the amount of energy it takes to extract a barrel of oil from them, not to mention the many wildlife habitats destroyed in the process.

Now take a look at it from the perspective of sustainable development. Currently, the global economy is driven by fossil fuels. Can any nation truly be expected to sit on the world's largest deposit of bitumen in that context and not try to benefit from it? At present, the oil sands produce one and a half million barrels of oil a day, and that number is expected to increase with time. Anybody who drives a car or flies in a plane is somewhat hypocritical when arguing against the development of these oil sands when those same individuals depend on the very product they produce. I also expect that given the chance to pay less at the gas pumps, most people would gladly do so without giving much thought as to how much more it would cost if the oil sands weren't being exploited.

Perhaps the argument could be made that only the cleaner sources of petroleum products should be sought and the dirtier ones left behind. But does that mean we should consider only the more pure oil deposits in the Middle East? Those are becoming depleted, and already the oil industry is looking to other untapped sources that are by nature going to be more difficult to extract and dirtier as a result. Like it or not, the days of easy oil acquisition are behind us. Offshore oil drilling is a perfect example of this. The British Petroleum disaster in the Gulf of Mexico in 2010 clearly shows that we can't rely on such drilling to be cleaner or better for the environment.

So where do we draw the line? Which sources of oil are okay to invest in, and which ones aren't? Until that has been defined on a global scale, I think it's reasonable to say that as long as the world is looking to petroleum to provide its energy, then a significant source such as the Athabasca oil sands is fair game. Thousands of Canadians have managed to get jobs in Alberta because of these reservoirs, and many of them have travelled from other parts of Canada where jobs are not as plentiful. The Canadian economy has benefited substantially from this resource, and both the United States and China are benefiting from it too. (The United States now imports more of its oil from Canada than it does from the Middle East.) Is it really fair to suggest that Canadians should leave the oil sands alone, losing thousands of jobs in the process and hurting the Canadian economy, when every other major nation on the planet is investing in trying to dig for their own fossil fuels or importing what they can't provide for themselves?

Until the time comes when the world is no longer dependent on fossil fuels, and individuals have reduced their personal carbon footprints to nearly zero, it's unrealistic to expect a source such as the oil sands to go untapped. (By the way, I am not advocating for the mining of these oil sands; obviously I would prefer to see Canada's fastest growing source of greenhouse gas emissions decrease but this isn't likely to occur until global attitudes toward fossil fuels change. Here, I am merely pointing out how sustainable development helps to shape such policies—for example, helping the Canadian economy with significant job creation and procuring an exportable resource currently needed all over the world at the cost of environmental harm and further greenhouse gas production. These sorts of decisions and policies are going to continue until every country has committed to making important changes. No one nation can be expected to hurt its own economy until all nations are prepared to take a similar approach. These are the reasons the Kyoto Protocol and the Copenhagen Accord have been largely disappointing; many nations believe that they're being targeted unfairly and that their economies are being asked to take a bigger hit.)

ARE WE GOING TO SOLVE THIS CRISIS OR NOT?

If I knew how to help people change behaviours so that they were able to face short-term pain more easily and receive the long-term gain, I would have many more ex-smokers in my practice. Just as it is with my patients and their unhealthy choices, I truly don't believe that there are any simple answers to the question of how to change the attitudes of the world with respect to greenhouse gas emissions. Despite the efforts of many knowledgeable and distinguished members of society, such as David Suzuki and Al Gore, real, meaningful change has yet to take place. Since I don't have the answers and don't pretend to, let me outline three possible scenarios that can occur from this point forward.

Scenario 1

We carry on with business as usual. If those who raise concerns about global warming and the climate crisis are wrong in their predictions—highly unlikely, but let's consider it for the moment—then change will be necessary only when we run out of fossil fuels, something that will occur at some point in the future. The skeptics claim that won't occur for many hundreds of years, and since they don't believe there is a climate crisis, that's why they argue against any changes now. Once those fossil fuels do run out, however, alternate sources of energy will have to be found. It's most likely that the world will turn to the renewable sources of energy discussed in chapter 7, not because people feel a moral obligation to live in cleaner and greener ways but simply because if they want to continue to live in the manner to which we've all become accustomed, energy will be needed from somewhere, and it will have to come from sources that won't run out so easily in order to be practical.

The same concerns raised about industries dependent on fossil fuels losing out during the transition will still be issues that society will have to contend with at that time, only it won't happen until the very last moment possible. If they don't adapt in the transition phase and ultimately become companies and businesses that are part of the new global economy, then they'll simply be left behind. But if the predictions of climate crisis and global warming are correct, then we'll face a different crisis long before we run out of fossil fuels. These are the same concerns addressed in chapter 5, and they have a real possibility of leading to the disruption and loss of millions if not billions of lives.

The people of our planet are very good at stepping up to help out in the face of a crisis on a grand scale, often putting aside party politics and international disputes in the name of helping our fellow men and women in their time of need. One only has to look to the tsunami in the Indian Ocean in

2004, which was particularly devastating for Thailand, leading to the deaths of more than 200,000 people. Individuals from all over the world donated funds to help deal with the disaster, rebuilding the country in the process. Another example is the more recent devastation experienced by Haiti from its earthquake in 2010. Again, the world offered to help regardless of policies, politics, or any perceived differences you can imagine. We saw people in need, and we offered to help. I have no doubt that if a crisis develops as a result of a changing climate, the world will once again step in to help. Of course, many will wonder why we had to wait for so many people to be affected by such devastation when the writing had been on the wall for so long.

This is one of the ironies in trying to get people to make a change today: there is no face of suffering at present that will help to get the message across because the bulk of the suffering will be in the future. (The best examples I've seen so far have involved stranded polar bears losing their Arctic ice; human examples are lacking, but given enough time, I'm sure we'll come up with some, unfortunately.) If there isn't an easily identified representative of the suffering, it's harder to care. In other words, out of sight, out of mind. Here is an interesting statistic that should give you pause: more time on CNN was devoted to the coverage of Baby Jessica, who in 1987 was trapped in a well in Texas for 58 hours before being rescued, than was given to all of the tragedies in Rwanda and Darfur combined, both of which lasted much longer.

This isn't simply an observation that has been made in media coverage; it has been demonstrated in studies on psychology. It turns out our empathy decreases as we're exposed to a greater number of sufferers. This fact has been put to good use by charitable organizations that advertise on television in their efforts to solicit donations: you are more likely to give money if they focus on one starving child in Africa or one solitary dog in an animal shelter than if they refer to the plight of the many others in the very same situation. As Joseph Stalin so aptly phrased it, "A single death is a tragedy; a million deaths is a statistic."

Scenario 2

We do our best to completely change what we're doing currently, as the most vocal of environmentalists urge. Since the writing is on the wall, and there's no time like the present, let's start changing things today. We're going to run out of fossil fuels and further hurt the atmosphere if we don't change. This would involve all nations coming together to put a price on carbon, capping the carbon dioxide we put into the atmosphere, and working together to create the infrastructure that will allow for such a dramatic change in global economy, allowing renewable sources to become the main focus rather than a minor asterisk or footnote.

The super grid will need to be created all over the world to accommodate the transportation of the energy that will be used. If we do make these sorts of changes, some industries will undoubtedly be hurt in this transition, as described earlier. And we'll be adding more carbon dioxide in the process since we'll have to use the energy sources of today to get us there. But of course, if we're successful in our endeavours, we'll hopefully avert major disasters on a global scale (if it isn't already too late, that is). We'll get our atmospheric carbon dioxide levels closer to what they were for many hundreds of thousands of years before we came along instead of maintaining the gradually increasing levels we've had over just the last two centuries since the Industrial Revolution began.

In that situation we'll have another irony to contend with: since we'll have prevented the catastrophes from occurring, there will always be skeptics who will claim that we didn't need to make those changes in the first place. They'll argue that this generation took too great of a sacrifice in their efforts to help the generations of the future. If you prevent a disaster, how can you truly ever be sure it was going to happen in the first place?

I often use an analogy in my practice that illustrates the principle of a disaster we manage to prevent. When a surgeon is talking to a patient after surgery, he or she can feel great pride that a life was just saved after a ruptured appendix has been removed, for example. But in my work as a cardiologist, most of the lifesaving I do is much less tangible. By keeping my patients on their aspirin and other medications to lower their blood pressure and cholesterol, I'm preventing many heart attacks, strokes, and deaths. But I never know who would have suffered one of these devastating outcomes, and neither do the patients. The lifesaving that medical rather than surgical doctors do is downstream rather than upfront and can never really be proven, at least not for individual patients. Our clinical trials have made it abundantly clear that the measures we take are the right ones, but individuals never know that they might have suffered a fatal heart attack were it not for the cardioprotective medications I have them on. (This is also why my surgical colleagues always get more bottles of wine from appreciative patients at the holidays than I do.)

Scenario 3

We decide to take a more balanced approach to sustainable development. We know we'll need to find a replacement for fossil fuels eventually since we know they won't last forever. And we know we'll have to do something to reduce the amount of carbon dioxide we put into the atmosphere. Converting to renewable sources of energy will help both problems and will benefit the planet because we won't wait until either is a crisis situation. Although there will always be skeptics who'll argue that change is unnecessary, we know it will be required someday even if climate were not part of the equation for the simple fact that our global economy currently depends on resources that aren't renewable. By taking the transition at a gentler pace, we can do our best to ensure that the principles of sustainable development are met—that is, ensure that we engage in "development that meets the needs of the present without compromising the ability of future generations to meet their own needs." We will have to be especially vigilant so that if we're incorrect in our predictions, we can modify our rate of change, minimizing any disasters arising from a pace that's too slow.

Only the passage of time will give us the answer as to which approach our species will ultimately take toward this issue. There will always be proponents on both sides of the spectrum, so that many will advocate for scenario 1, while their polar opposites advocate for scenario 2. I'm not sure which path is truly the best—although I certainly discount scenario 1 as a viable consideration—but I predict that we're destined to follow the path of scenario 3 because it's the closest thing we have to a compromise. Certainly, there are changes in place that indicate we're doing something—we recycle, we drive more hybrids than we used to, and we seem more aware and more conscientious of these issues compared to just a generation ago—but we aren't moving nearly as quickly as we could if we really needed to, if we were already facing a crisis situation now rather than sometime in the future.

But it's important to remember that our world is made up of many people with different ideals, goals, aspirations, beliefs, and attitudes. This planet belongs to all of us. A more centrist and compromising approach seems to be most likely, given our past history as a civilization. But is that going to be enough? Given the facts you now possess, I'll let you decide for yourself which approach you think is best. As to whether we will all choose wisely in the manner we tackle this problem as a species, only time will tell.

Key Concepts of Chapter 8

- The concept of utility quotients helps to explain why changing attitudes is difficult despite the long-term benefits that can be gained.
- Sustainable development is important to ensure that no one group suffers unfairly when considering changes in policy.
- The most important choices to help solve the problems at hand are up to each and every one of us.

Suggested Reading for Chapter 8

Ariely, Dan. *The Upside of Irrationality: The Unexpected Benefits of Defying Logic at Work and at Home.* New York, NY: HarperCollins, 2010.

Gladwell, Malcolm. *Blink: The Power of Thinking without Thinking.* New York, NY: Little, Brown, 2007.

Harris, Sam. *The Moral Landscape: How Science Can Determine Human Values.* New York, NY: Free Press, 2010.

Epilogue

> There are no passengers
> on Spaceship Earth.
> We are all crew.
> —Marshall McLuhan

One of the biggest difficulties we face as a civilization is the inability to perceive our place in Earth's history. We have a very limited perspective simply because our personal experiences are restricted to the life span of a human being. We marvel at progress while looking back to decades past and express amazement at all of the changes the world has undergone, but we struggle with time scales that are much grander than one solitary lifetime.

When I reflect on my life so far, which started in the 1960s, I'm as amazed as anyone at the various transformations that have taken place. I've watched old buildings come down and new buildings go up, television go from black and white without cable to high-definition wide-screen digital images. The act of writing and mailing letters and waiting weeks for a response has been replaced with e-mailing and the even more instantaneous text messaging. Listening to music no longer requires a turntable, and in fact, now I have my entire music library digitally recorded on a handheld device that is smaller than my wallet. I managed to get through my education up to and including all of my university years without ever owning a computer, but now computers and PDAs make up a vital component of the learning process, and no student can be without them.

And just like every other parent out there, my children laugh at me when I try to explain how different things were for me in the "old days." All of this has taken place in the span of mere decades. But despite how much I love and study history, I truly have a hard time "feeling" the years beyond my own lifetime. I believe most of you probably have the same experiences. If it's this difficult to relate to the past when so much of history is recorded, think of how much more difficult it is to relate beyond our years to a future when we won't even be here.

Since we can easily appreciate the scales of time that are within our personal experience, I thought it would be a useful exercise to take the 4.5-billion-year history of our planet and shrink it down to the span of just one year. In this new timeline, the Earth is formed out of the residual material that resulted from an earlier supernova on New Year's Day, January 1. The earliest life forms don't pop into existence until the middle of February, and those are limited to simple prokaryotes such as bacteria, those cellular life forms without nuclei. The more complex eukaryotes, which include plant and animal life, don't come along until September, and by that point, we're already more than two-thirds through our planet's entire history. Mammals (of which you and I are examples) don't show up until the middle of December. The dinosaurs that were wiped out at least in part because of the devastating effects of a large asteroid 65 million years ago don't disappear until December 26, with less than a week to go before this compressed year is over. The species *Homo sapiens*—better known as human beings—finally makes its mark, but not until New Year's Eve, December 31, around 9:00 p.m. The Industrial Revolution that changed the world so dramatically started with only two seconds left on the clock to go before midnight. People who manage to live very long lives, far beyond a typical life span, still live for mere tenths of a second on this scale. All of the people who believe that we're masters of this planet's destiny simply because we've managed to alter it so dramatically in such a short period of time demonstrate a lack of respect for everything that came before us, all of which helped to get us where we are today. Such an attitude also displays a lack of responsibility to future generations, who deserve not only the same kind of existence that we have enjoyed but in fact a better one.

Put into this context, it might be a little easier to consider how we live our lives and the consequences of the actions we take with better perspective. We need to be more conscientious of how these tenths of a second in which we live out our lives in our condensed Earth-year are going to impact on the next few seconds in our planet's history. Can we really cast a blind eye to the entire record of Earth's heritage and potentially cause irreparable harm in what amounts to the blink of an eye, simply because we lack the perspicacity to see our planet's history in its entirety? This will very likely prove to be an ongoing struggle for modern society. Hopefully, it won't also prove to be our undoing.

This book began with a description of a famous picture of Earth taken by the *Apollo 17* astronauts, showing our beautiful planet in all its glory and wonder. Now that you've reached the end of this book, there's another picture of our home that I'd like to describe. It isn't as famous as the first one, but I think its significance and impact are far greater.

On September 5, 1977, *Voyager 1* was launched from Cape Canaveral in Florida. Its mission was to take images of the two largest planets in our solar system, namely Jupiter and Saturn. Because the spacecraft would be so much closer to them during its journey than the Earth-based telescopes we'd been using, the portraits of these planetary neighbours would be far superior to anything we'd ever seen to that point.

It was also known that in the coming years Jupiter, Saturn, Uranus, and Neptune would be conveniently lined up so that a spacecraft could get close enough to each one in a single trip, allowing for a level of study that was previously unavailable. *Voyager 2* was launched the same year and followed a slightly different trajectory that allowed it to study all four of the outermost planets. The information these two man-made explorers gathered by working in tandem proved invaluable to the studies of our solar system. The proximity to each planet in turn allowed for some of the most stunning astronomical photos ever taken. You've likely seen many of these pictures because they are still the best photographic depictions we have of these gas giants that lie beyond the asteroid belt.

It was an incredibly long trip that took quite some time: *Voyager 2* reached Jupiter in 1979, Saturn in 1981, Uranus in 1986, and Neptune in 1989. These two spacecraft continue to move farther away from us each second, and *Voyager 1* is the most distant object from Earth ever made by the human species, currently almost 17 billion kilometres or 11 billion miles away, about 115 times the distance between the sun and the Earth.

Once *Voyager 1* had fulfilled its mission, the brilliant astronomer Carl Sagan asked NASA officials if they would have the camera look back and photograph the solar system from an outside perspective. Since there wasn't much more planned for the spacecraft—and a request from Carl Sagan, who had helped to popularize astronomy better than anyone, was not something to be taken lightly—they readily agreed. Through communications back and forth using radio waves, in 1990 a number of images were obtained of the various planets, none of which were as eye-catching as the beautiful photographs already taken during the fly-bys. But when a photograph of Earth was obtained, it proved to be the most impressive of all. This wasn't because it displayed some brilliant vista that would make you stand up and take notice like the photograph taken by the Apollo astronauts. Rather, it was impressive because of how humble it made the viewer feel when he or she looked at it.

In chapter 1, I mentioned that Earth is often referred to as a big blue marble in photos of it taken from space, but from this great distance—more than six billion kilometres or four billion miles away—it was nothing more than a pale blue dot, barely noticeable, suspended within a beam of sunlight that was present in the optics of the camera due to the small angle between the Earth and the sun from that distance. Nothing but a pale blue dot; no discernible features such as oceans, clouds, or land masses can be identified in the image. For something so surprisingly minimal in pixel count, its significance to everyone who sees it is beyond description. It has found its way all over the Internet and has been acknowledged as one of the top ten space images ever taken.

Dr. Sagan was at least as impressed as anyone else who gazed at this image. Sagan titled his 1994 book *Pale Blue Dot* for this very reason. In the first chapter of that book, titled "You Are Here," he described this image and the perspectives it can invoke more eloquently than anyone I can imagine. Because no summary could possibly do his words justice, I have obtained permission from his estate to reproduce them here.

Our planet from more than six billion kilometres away, "the only home we've ever known." (NASA/JPL)

From this distant vantage point, the Earth might not seem of particular interest. But for us, it's different. Look again at that dot. That's here, that's home, that's us. On it everyone you love, everyone you know, everyone you ever heard of, every human being who ever was, lived out their lives. The aggregate of our joy and suffering, thousands of confident religions, ideologies, and economic doctrines, every hunter and forager, every hero and coward, every creator and destroyer of civilization, every king and peasant, every young couple in love, every mother and father, hopeful child, inventor and explorer, every teacher of morals, every corrupt politician, every "superstar," every "supreme leader," every saint and sinner in the history of our species lived there—on a mote of dust suspended in a sunbeam.

> The Earth is a very small stage in a vast cosmic arena. Think of the rivers of blood spilled by all those generals and emperors so that, in glory and triumph, they could become the momentary masters of a fraction of a dot. Think of the endless cruelties visited by the inhabitants of one corner of this pixel on the scarcely distinguishable inhabitants of some other corner, how frequent their misunderstandings, how eager they are to kill one another, how fervent their hatreds.
>
> Our posturings, our imagined self-importance, the delusion that we have some privileged position in the Universe, are challenged by this point of pale light. Our planet is a lonely speck in the great enveloping cosmic dark. In our obscurity, in all this vastness, there is no hint that help will come from elsewhere to save us from ourselves.
>
> The Earth is the only world known so far to harbor life. There is nowhere else, at least in the near future, to which our species could migrate. Visit, yes. Settle, not yet. Like it or not, for the moment the Earth is where we make our stand.
>
> It has been said that astronomy is a humbling and character-building experience. There is perhaps no better demonstration of the folly of human conceits than this distant image of our tiny world. To me, it underscores our responsibility to deal more kindly with one another, and to preserve and cherish the pale blue dot, the only home we've ever known.[1]

There is no way that I could hope to write something that would put it into a more proper perspective or to motivate you more to care for your home planet than those beautifully written words. Hopefully, you have gained something from reading this book. At the very least, you have likely learned a few things on the various topics that pertain to global warming and the climate crisis. Perhaps, especially with the help of Dr. Sagan's words, you will feel a certain responsibility to do your part to make a difference. We are nothing more than caretakers of our world for the short time that we live here. As such, we must ensure that we leave it in a state where our children

---

1 Carl Sagan, *Pale Blue Dot: A Vision of the Human Future in Space* (New York, NY: Random House, 1994), 8–9.

and grandchildren can enjoy it as well as we have. Remember the proverb: you don't inherit the planet from your parents; you borrow it from your children.

Thank you for reading, and I wish you the best of success in your personal efforts to help save our planet.

Acknowledgments

There are many people I need to thank who helped me through the process of getting my book from its initial stages, when it was nothing more than an idea, to the object that you're now holding in your hands.

Patrick Brown, the member of Parliament where I work in Barrie, Ontario, has always been supportive of projects in which I've been involved. When I told him that I cared about the environment and was interested in doing something to help educate the public about it, he quickly put me in touch with the then federal minister of the environment, the Hon. Jim Prentice. Having the opportunity to discuss issues with the environment minister over a few hours helped to steer me in the direction I ultimately took, which was to write this book. Along the way, Mr. Prentice also appointed me to his Sustainable Development Advisory Council, which taught me even more about the issues at hand and helped to put me in contact with many individuals who were much more knowledgeable about the environment than I was. I would particularly like to thank Mr. Peter Robinson, the CEO from the David Suzuki Foundation, who was encouraging in his communications with me and also agreed to review an early version of my manuscript despite his very busy schedule.

I also appreciate everyone who helped by looking at various drafts of the manuscript and then providing me with excellent advice, ultimately improving the final product greatly. Dorothy and Allan Sarjeant provided me with some very helpful comments. Dr. Robert Ballagh and Margot Douglas, two of the best friends that anyone could hope for, were particularly helpful and supportive in so many ways that they will always have my undying gratitude. Dr. Ballagh's artistic skills were also put to good use when I asked if he could provide me the illustration I used for nuclear fission. Drs. Monique Christakis, Tej Deol, and Marc Lewin were extremely encouraging when I crossed their paths at my twenty-year medical school reunion, and their support really helped to give me the final push I needed to get my book completed.

I would be remiss if I didn't thank my parents. The education they helped to provide for me and the encouragement I received growing up in pursuing my goals has always stayed with me, and I wouldn't be the person I am today if not for their love and support.

Although writing isn't my profession, it has always been something I love to do, and there are some individuals who helped me to hone my skills. Most importantly among these was the late Dr. Sylvia Bowerbank who taught me greatly when I took her writing course at McMaster University in Hamilton, Ontario, before I pursued my medical education. I will never forget the interesting discussions we shared on what defines "style." To this day, I still do my best to achieve good marks from her with everything I write, and that is largely thanks to her guidance and instruction.

To hope to be a good writer requires emulating great writers, and there are a few I would like to single out. John Irving has always been my favourite novelist, and reading his words over many years and seeing him interviewed when he has discussed his writing process have helped me to improve. He taught me that it isn't as important to excel at writing so much as it is to excel at rewriting. Other authors who have been influential to me include Bill Bryson and Stephen King. (I highly recommend Mr. King's book *On Writing*, which is both enlightening and entertaining, even if you have no aspirations to put pen to page.)

The most influential writer and scientist in my life has always been the late Dr. Carl Sagan. He shaped my interest and love for science when I first began to read his work while I was in high school, ultimately leading me to my chosen career in medicine. Many times when I felt stuck at various points in writing this book, I was able to get myself out of it by simply asking, "How would he write about this?" When it came time to end the book, I thought his words were best suited for the task. His estate was extremely helpful and encouraging when I requested permission to reproduce a few paragraphs, and I couldn't be more thrilled to have been able to interact with them. His wife,

Ms. Ann Druyan, was very considerate in giving me this permission, and I owe her my gratitude as well.

Members of the editorial staff at iUniverse were very helpful in getting me to my final product, and as much as criticism can be tough to take, when it's constructive and helps you to improve your work, it's not only worthwhile but also necessary.

Finally, without the love and support of my wife, Katherine, and my two sons, Matthew and Jamie, none of this ever would have been possible. The boys now understand the issues described in this book better than I would have ever hoped for. (Jamie's placement of "Save the Earth" stickers on our refrigerator and the 100 percent Matthew scored on the climate crisis exam I gave him are proof of that!) Katherine did everything possible to make this project of mine a reality. As a writer herself, she helped me with the manuscript enough to produce something that was both readable and educational. She then changed hats and became my first editor, getting rid of words I like to overuse ("actually") and suggesting modifications that only enhanced the concepts I was trying to get across. My family is my rock, and it's for this reason that I've dedicated this book to them.

Glossary

aerobic. Something occurring in the presence of oxygen. An example of an aerobic process is respiration.

albedo. The amount of light reflected from Earth's surface. The higher the albedo, the more reflective the surface is.

alternating current. A type of electricity that is best adapted for transmission across vast distances through power lines for distribution to homes and businesses.

anaerobic. Something occurring in the absence of oxygen. Anaerobic chemical reactions often produce different products than aerobic chemical reactions.

anthracite. The highest grade of coal, due to its higher carbon content.

aphelion. The point in a planet's orbit when it is farthest away from the sun.

asteroid. A rocky body found within our solar system that orbits the sun. Most asteroids in our solar system are found within the asteroid belt located between Mars and Jupiter.

atom. The most basic unit of a chemical element. The building blocks of atoms are protons, neutrons, and electrons.

atomic mass. The mass of an atom, comprising the protons and neutrons found within its nucleus.

atomic number. The number of protons contained within the nucleus of an atom. It defines the element to which an atom belongs.

big bang theory. The theory of the origin of our universe where a massive explosion created all the matter found within it nearly fourteen billion years ago.

bituminous coal. An intermediate grade of coal that is higher in quality than lignite but lower than anthracite. It contains a tarry substance known as bitumen.

brine rejection. The process where freezing salt water in the oceans expels its salt, leading to ice that is much closer to fresh water in content.

carbohydrate. An organic molecule made up of carbon, hydrogen, and oxygen atoms. Carbohydrates can store energy and are used by life on Earth as food.

carbon cycle. The natural circulation of carbon on Earth through various processes, including photosynthesis, combustion, volcanism, and the movement of tectonic plates.

carbon offsets. A way to reduce global greenhouse gas emissions by investing in the development of measures that reduce emissions elsewhere, offsetting those that are added locally.

chlorophyll. The green pigment found in plants where the chemical process of photosynthesis occurs.

comet. A body found in our solar system that is typically composed of minerals and ice. Bombardments by comets in Earth's early history may have provided our planet with most of its water.

concentrated solar thermal. A form of solar energy where mirrors reflect sunlight onto a tower of water, boiling the water into steam, which is used in a turbine to generate electricity.

Copenhagen Accord. A document that was created at a meeting of the nations of the world held in Copenhagen, Denmark, in 2009. It addressed the ongoing need to further tackle global warming.

deforestation. The process of removing trees and other plant life to make room so that the land can be used for other purposes. Frequently, this is done by burning forests rather than simply cutting them down.

direct current. A type of electric current that will degrade if transmitted across great distances. Batteries provide direct current.

disaccharide. A type of sugar that is composed of two monosaccharide units. Sucrose (common table sugar) is a disaccharide consisting of one glucose molecule and one fructose molecule bonded together.

eccentricity. The measure of degree of variation in an ellipse. An eccentricity of zero is a perfect circle; the closer the value is to one, the flatter the ellipse is. The orbit of Earth demonstrates some eccentricity and affects climate naturally over the long-term.

electricity. Energy resulting from the movement of charged particles, such as electrons. Such movement is called an electric current.

electromagnetic spectrum. The range of wavelengths over which electromagnetic radiation is found. It includes visible light, gamma rays, X-rays, infrared radiation, ultraviolet light, microwaves, and radio waves.

electron. A subatomic particle that orbits the nucleus. Electrons have a negative charge equal to the positive charge of the proton found within the nucleus and almost no mass, so they do not contribute to the atomic weight.

eukaryotes. Organisms that contain cells that have DNA within a nucleus. They are more complex than prokaryotes and comprise all life forms on Earth that are more evolved than bacteria.

fission. The nuclear reaction where an element splits into two or more elements of smaller mass. It is associated with a significant release of energy and is a source of nuclear power.

Framework Convention on Climate Change (FCCC). Created by the United Nations at the Earth Summit in 1992, it was the first international effort to assess the problem of greenhouse gases. It continues to be active.

fusion. A nuclear reaction where two atoms combine together to produce an atom of a greater atomic number. It is associated with a significant release of energy and powers the sun.

gas giants. The larger planets contained in the outer portion of our solar system, which are composed of massive amounts of gas rather than having a rocky crust for the surface. Jupiter, Saturn, Uranus, and Neptune are all gas giants.

geoexchange. A process to heat and cool buildings by using a ground source heat pump, which pumps heat to and from the ground below the frost line.

geothermal. A method of obtaining latent heat from deep inside the Earth as a source of energy.

gigaton. A billion tons.

greenhouse gas. A gas found within Earth's atmosphere that allows visible light from the sun through but reflects infrared radiation back, effectively acting as insulation.

Gulf Stream. A natural conveyor belt that exists in the Atlantic Ocean. It carries heat from the Caribbean Sea to northwestern Europe, primarily the United Kingdom and Scandinavia.

hydrocarbons. Chemical compounds that contain hydrogen and carbon and are often used as sources of fuel.

Industrial Revolution. The modernization of industry with the help of machinery that started in the eighteenth century as a result of the development of the steam engine.

infrared radiation. The part of the electromagnetic spectrum that is below red in the visible spectrum and radiates energy in the form of heat.

irradiance. The amount of sunlight produced by the sun.

isotope. A different form of an element where the nucleus contains the same number of protons but has a different number of neutrons, leading to a different atomic mass.

Kyoto Protocol. A treaty written up in 1989 in Kyoto, Japan, which addressed the need for the developed nations of the world to reduce greenhouse gas emissions.

lignite. Often referred to as brown coal, it's the lowest grade of coal that developed from peat. It can be an intermediate step to the formation of higher grades of coal, such as bituminous coal and anthracite.

meiosis. A type of cell division that produces two daughter cells, each containing exactly half of the genetic information of the parent cell. Meiosis is used to produce eggs and sperm cells.

mitosis. A type of cell division that produces two daughter cells that are identical to the parent cell. Most cells in our bodies replace themselves using mitosis.

molecule. A group of atoms held together by chemical bonds. A molecule is the smallest fundamental form of a substance that can participate in chemical reactions.

monosaccharide. A class of sugar molecules that cannot be split into simpler sugars. Glucose is a monosaccharide.

Montreal Protocol. A treaty written up in 1987 in Montreal, Canada. It addressed the need for a reduction in the use of chlorofluorocarbons, which are harmful to the Earth's ozone layer.

neutron. A particle that is found within the nucleus of an atom. The number of protons and neutrons in the nucleus together defines the atom's atomic weight. The neutron has no charge associated with it but provides stability to the atom.

nuclear reactor. A power station that generates electricity by using energy obtained through the fission of radioactive materials. The energy is used to boil water into steam and rotate a turbine.

nucleus. The centre of an atom where the protons and neutrons are contained. Also the centre of a cell found in plants and animals where DNA is located.

orbital forcings. Changes in Earth's orbit, such as tilt and eccentricity, that naturally affect climate over a 100,000-year cycle.

oxidation reaction. A chemical reaction where an ion, atom, or molecule loses electrons, often to an oxygen atom, with an oxide formed in the process.

ozone. A molecule made up of three atoms of oxygen. A layer of ozone in the upper atmosphere protects Earth's surface from the harmful effects of ultraviolet radiation.

perihelion. That point in a planet's orbit when it is closest to the sun.

periodicity. The repetition of certain properties in chemical elements in a periodic pattern. This enables the elements to be positioned in an organized fashion in the periodic table.

periodic table. A table listing all of the known elements in an organized fashion, with certain common features demonstrated among various groupings.

photon. A particle of light representing electromagnetic radiation.

photosynthesis. A chemical process whereby the chlorophyll in plants can manufacture carbohydrates out of water and carbon dioxide using sunlight as its energy source.

photovoltaic panels. Also known as solar panels, they produce electricity when photons hit their surface and excite electrons, initiating an electric current.

planetesimals. Rocky bodies found within a solar system that are formed by the aggregation of smaller bodies, such as asteroids. If they continue to grow due to further collisions, they may become planets.

polysaccharides. Carbohydrates that are composed of many individual sugar molecules. Examples include starch, cellulose, and glycogen.

prokaryotes. Simple single-celled organisms that are among some of the first life forms that evolved on Earth. All bacteria are prokaryotes.

proton. A particle found within the nucleus of an atom. The number of protons defines an atom's atomic number and therefore the element to which it belongs. A proton has a positive charge equal to but opposite from the electron.

reduction reaction. A chemical reaction where an ion, atom, or molecule gains electrons in the process. Reduction-oxygenation reactions generally occur together and are called redox reactions for short.

regenerative braking. A process used in hybrid vehicles where the battery is charged while coasting and braking, allowing for less gasoline to be used because the battery contributes to the vehicle's power.

smart grid. An advanced method of electricity distribution using two-way digital technology.

steam engine. An invention improved by James Watt in the eighteenth century. By using steam, it was able to harness power and was useful in mills and locomotives. Its popularity helped to herald the Industrial Revolution.

subduction. The process where one tectonic plate can move under another, arising from a collision between them, burying materials from the surface of one plate deep underground as a result.

super grid. An advanced technology that allows for the transmission of much higher voltages across greater distances compared with the energy grids used today.

supernova. One way a star can reach the end of its life. As a result of the massive explosion associated with it, many elements of higher atomic numbers are formed through the process of fusion.

sustainable development. A policy of ensuring that the needs of future generations are met without compromising the needs of the present generation.

tectonic plates. Components of the Earth's crust that float on the liquid magma beneath them. They are mobile and have created many geological features found on our planet.

turbine. A machine that generates electricity, whereby rotational energy caused by something moving through it or pushing on pistons connected to it, such as water, wind, or steam, produces an electric current.

ultraviolet radiation. The part of the electromagnetic spectrum that is above violet. It is a form of energy that we cannot see, but its effects are noted with suntans and sunburns.

United Nations Framework Convention on Climate Change. See Framework Convention on Climate Change.

vector. An organism that transmits a disease from one plant or animal to another.

List of Illustrations

1. Earth's atmosphere visible from space.
2. A helium atom consisting of two protons and two neutrons surrounded by two orbiting electrons.
3. The periodic table.
4. The Himalayas were created by the slow collision of two tectonic plates.
5. The steam engine that led to the Industrial Revolution.
6. A molecule of methane.
7. A molecule of butane.
8. Sound waves expand just like ripples in a pond.
9. Sir Isaac Newton (1642–1727), English scientist and mathematician, using a prism to break white light into its spectrum. With Cambridge roommate John Wickins. Engraving of 1874.
10. An example of deforestation.
11. Carbon dioxide levels fluctuate naturally throughout the year.
12. Most of Earth's land mass is located in the Northern Hemisphere.
13. The Keeling Curve, showing that measurements of atmospheric carbon dioxide have experienced a steady increase over the years.
14. Global warming will have devastating consequences for the environment.
15. A hurricane as seen from space.
16. A system of geoexchange used to heat and air-condition a home.

17. An array of solar panels "farming" the sun's energy.
18. Mirrors reflecting sunlight to the top of a concentrated solar thermal tower located in Spain.
19. A wind farm located in Whitewater, California.
20. Although nuclear energy avoids the problem of greenhouse gas emissions, uranium mining and nuclear waste have their own environmental concerns.
21. An example of nuclear fission: a neutron collides with a uranium atom, producing two smaller atoms and releasing energy in the process; three neutrons are also released, allowing the reaction to continue.
22. The Athabasca Tar Sands in Alberta, Canada, contain large amounts of bitumen that can be converted into oil.
23. Our planet from more than six billion kilometres away, "the only home we've ever known."

Index

A

Acid Rain Treaty, 144
acidity, 88–89, 106
activism, 122
aerobic process, 179
aerobic respiration, 26–27
agriculture, 70, 71
albedo, 84–85, 179
algae, 25, 34–35, 139–140
algal fuels, 139–140
alternating current (AC), 117, 129, 179
An Inconvenient Truth, ix, 84
anaerobic, 33–34, 179
Annex 1 nations, 145
Antarctic, 73, 77, 82, 84, 86
anthracite, 34, 53, 179
aphelion, 97, 179
Apollo 17, 3–5, 169
appliances, 108
Arctic, 84, 86, 92
argon, 67, 69
asteroids, 28–29, 179
Athabasca oil sands, 115–116, 125, 157–159
atherosclerosis, 150
atmosphere, 5–6, 67–69
atomic mass, 179
atomic number, 7–8, 180
atoms, 7–8, 17, 52–53, 179
aurora australis, 5–6
aurora borealis, 5

B

bacteria, 29–30, 168, 185
basicity, 88
batteries, 48–49, 117, 181
Bentham, Jeremy, 153
beta decay, 137
big bang theory, 9–10, 180
biodiesel, 139
biofuels, 138–140
bitumen, 157–158
bituminous coal, 34, 180
brine rejection, 92, 180
Bullfrog Power, 121
butane, 54–55
buying locally, 110–111

C

cable television, 127–128
cap-and-trade system, 146–147
carbohydrates, 25, 32, 180
carbon
    distribution of, 17–19, 27
    in form of coal, 53
    fusion and, 13–14
    origin of, 18
carbon capture and storage, 140–142
carbon cycle
    described, 21, 51, 180
    before human activity, 27–32
    on land, 32–34
    in water, 34–36
carbon dating, 81–82, 137

carbon dioxide
    annual cycle of, 74–75
    carbon capture and, 140–142
    carbonated beverages and, 35
    concentrations of, 32, 67, 69
    in earth's history, 28
    formation of, 53–56
    increasing levels of, 51, 72–77
    non-fossil fuel sources of, 70–72
    in oceans, 87–88
    reducing levels of, 105–106, 140–142
    See also greenhouse gases
carbon footprint calculators, 114–115
carbon offsets, 114–115, 180
carbonated beverages, 35
carbonic acid, 35, 87–88
cardiovascular disease, 150–151
cars. *See* vehicles
cattle, 71–72, 139
cellulose, 25, 27, 71, 185
cerebrovascular accidents, 150
change, 149–153
chemical bonds, 52–53, 135
chemical reactions, 24–26, 48–49, 52–53
chlorofluorocarbons (CFCs), 64, 69, 144, 183
chlorophyll, 25, 63, 180
cigarette smoking, 151–153
claudication, 150
climate change
    benefits, 84
    natural causes of, 95–98
    severe weather, 89–93
    tipping point for, 105
    See also global warming
coal
    combustion of, 53–56
    as fossil fuel, 50–51
    grades of, 34
    mining sources of, 126
    steam engine and, 46
coalification, 34
coastal flooding, 87
colours, 62–63
comets, 28–29, 180
Commonwealth Scientific and Industrial Research Organisation (CSIRO), 86
compact fluorescent light bulbs (CFLs), 108
composting, 109–110
computed tomography, 65

concentrated solar thermal (CST), 130, 131, 180
conservation of mass, 25–26
consumption, 107–112
control rods, 137
Copenhagen Accord, 146, 159, 180
coral reefs, 89
costs, 95, 107, 120, 121
cows, 71–72
crude oil, 38, 50–51
CT scans, 65

D

Darwin, Charles, 31
decisions, 152–153
deforestation, 70–71, 181
deuterium, 81
diamonds, 17, 18
direct current (DC), 117, 129, 181
disaccharide, 181
disease, increased spread of, 93–95
downsizing, 112
droughts, 91

E

$E=mc^2$ equation, 12–13, 134
Earth photos, 3–5, 169–171
earthquakes, 37
eccentricity, 97–98, 181
efficiency, improved, 107–112
Einstein, Albert, 12
El Niño-Southern Oscillation, 96
electric motor, 49–50
electric vehicles, 128–129
electricity
    defined, 48, 181
    generation of, 48–49
    greener options, 121
    magnetism and, 49–50
    regulating and storing of, 126–128
    See also power stations
electromagnetic (EM) spectrum, 49, 60–65, 116, 181
electrons, 7–8, 181
elements, 6–9, 9–17
emissions trading, 147
empathy, 161
encephalitis, 95

energy audits, 108
energy sources
    criteria for, 43–44, 124
    fossil fuels as, 44–45
    in the future, 126–128
    historical, 45–46
    renewable, 115–122, 129–140, 163
    wood-burning, 70, 138
energy-efficient vehicles, 112–113
enhanced geothermal energy, 133
enhanced oil recovery, 141
ethane, 54
ethanol, 139
eukaryotes, 30–31, 168, 181
evolution, 31

F

Faraday, Michael, 49–50
farming, 70, 71
financial incentives, 142–143
fission, 12, 134–137, 181
floods, 87, 91
forest fires, 70
fossil fuels
    benefits of, 44–45
    combustion of, 72–74
    continued use of, 140
    differences in, 52–57
    efficiencies of, 50–51
    finite supply of, 124–125
    global economy and, 144–145
    origin of, 38–39
    remote locations of, 126
    steam engine and, 46
Framework Convention on Climate Change (FCCC), 145–146, 181
Franklin, Benjamin, 48
Frito Lay, 111–112
fusion, 11–14, 15, 26, 182

G

Galvani, Luigi, 48
gamma rays, 65
garbage, 109–110
gas giants, 16, 182
geoexchange process, 118–120, 182
geothermal energy, 118–120, 133–134, 182
gigaton, 182

Gillett, Nathan, 105
glaciers, 84, 85
global efforts, 144–147
global population, 72
global warming
    increasing temperatures, 90–91
    natural causes of, 95–98
    potential benefits of, 84
    temperature measurements and, 80–83
    tipping point for, 105
    winter storms and, 91, 155
    See also greenhouse effect
glucose, 25–27, 32, 33, 183
glycogen, 33, 185
gold, 17
Gore, Al, ix, 84, 160
governmental policies, 142–144
Great Britain, 92
greenhouse effect, 59, 65–67
greenhouse gases
    concentrations of, 67–69
    defined, 182
    insulating effects of, 67, 79–80, 85
    other sources of, 70–72
    putting a price on, 141–142
    See also carbon dioxide; methane
Grinnell Glacier, 85
ground source heat pump, 118–120
Gulf Stream, 92, 182

H

half-life, 81–82
Hawaii, 73
heart attacks, 150–151
heavy water, 23–24, 81
helium, 8, 11–14, 15
Himalayas, 36–37
hurricanes, 89–90
hybrid vehicles, 113
hydrocarbons, 54–55, 182
hydrogen, 8, 10–12, 15
hydrothermal plants, 133

I

ice core samples, 76–77, 80, 82
incentives, 142–143
Industrial Revolution, 46–47, 182
infrared radiation, 64–65, 66–67, 84–85, 182

international treaties, 144–147
iron, 14, 16
irradiance, 96, 98, 182
isotopes, 80–81, 183

K

Keeling, Charles David, 73, 76, 104
Keeling Curve, 75–76
kinetic energy, 12–13
Kyoto Protocol, 145–147, 159, 183

L

lifestyle changes, 150–153
light, speed of, 12–13, 62
light bulbs, 108
light energy, 62–65, 116
light-emitting diodes (LEDs), 108
lignite, 34, 183
Little Ice Age, 96
livestock, 71–72
lobbyists, 143
Loblaws, 111
locavores, 110
Lyme disease, 93, 95

M

magnetism, 49–50
malaria, 93–94
Mauna Loa, 73
meat consumption, 111
Medieval Warm Period, 96
meiosis, 30, 183
melting ice, 84–86, 92
Mendeleev, Dmitri, 8–9
methane, 38, 50, 54–56, 67, 69, 71–72, 135, 139
microwave radiation, 65
Mill, John Stuart, 153
Miller-Urey experiment, 29
mitosis, 30, 183
molecules, 10, 183
monosaccharide, 183
Montreal Protocol, 144, 183
mosquitos, 93–94

N

natural gas, 38, 50–51, 52, 54–56, 121
natural selection, 31
neutrons, 7–8, 183
Newton, Isaac, 63, 64
nitrogen, 67, 68
nitrous oxide, 69, 77
noble gases, 69
North Atlantic Oscillation, 96
north star, 97
northern lights, 5
nuclear bonds, 53, 134–135
nuclear fusion, 11–12, 26, 182
nuclear power, 50, 134–138
nuclear reactors, 137, 183
nuclear waste, 138
nucleus, 7–8, 30, 184

O

obliquity, 97
oceans
    carbon dioxide in, 87–88, 106
    increased acidity of, 88–89, 106
    rising levels of, 86–87
octane, 55, 56
offshore drilling, 115–116, 124–125, 158
oil, 38, 50–51
oil sands, 115–116, 125, 157–159
Olmedilla Photovoltaic Park, 129, 130
orbital forcings, 96–98, 184
Orsted, Hans Christian, 49–50
oscillations, 96, 98
oxidation reaction, 48–49, 184
oxygen
    in atmosphere, 32, 67, 68
    and changing ratio of isotopes, 82
    cycle of an oxygen atom, 18, 24–27
    life depends on, 6
ozone, 64, 69, 144, 183, 184

P

paleoclimatology, 80
paraffin wax, 55
particle accelerators, 7, 15, 17
peat, 33–34, 139
perihelion, 97, 184
periodic table, 8–9, 13, 184

periodicity, 8, 184
petroleum, 38, 50–51
pH scale, 88
photosynthesis
    combustion of fossil fuels and, 73, 74
    creation of oxygen by, 6, 31–32, 63
    defined, 184
    in a greenhouse, 66
    oxygen atoms and, 25–26
    in response to increased carbon dioxide, 106
    in water, 34–36
    wood-burning and, 138–139
photovoltaic panels, 116, 129, 184
planetesimals, 16, 184
plant life, 31–32, 66–67
Pluto, 6
plutonium, 15, 137
polar bears, 92, 161
polar ice caps, 84–86, 92
polysaccharides, 185
population, global, 72
power grids. *See* smart grid; super grid
power stations, 126–128
precession, 97–98
precycling, 109
primary prevention, 151
prisms, 62, 63, 64–65
programmable thermostat, 109
prokaryotes, 29–30, 168, 185
propane, 54
protons, 7–8, 185
PS10 Power Tower, 130

Q

quarks, 7

R

radio waves, 65
radioactive decay, 81–82, 137
radioactive waste, 138
recycling, 109–110
redox reactions, 48–49, 185
reduction reactions, 48–49, 185
refrigeration, 118–119
regenerative braking, 113, 185
resistance
    to fighting climate change, 154–156
    to healthier lifestyles, 149–153
respiration, 26–27
Revelle, Roger, 73, 76, 104–105
Ring of Fire, 133
rising sea levels, 86–87
Rocky Mountain spotted fever, 95
ruminants, 71–72, 139
rust, 48–49

S

Sagan, Carl, 17, 169–172
San Andreas fault, 37
Scandinavia, 92
Schumacher, E. F., 112
Scripps CO_2 program, 69, 73–74
scrubbing, 141
secondary prevention, 151
severe weather, 89–93
sheep, 71–72, 139
smart grid, 127–128, 143, 185
smoking, 151–153
solar energy, 116–117, 129–130
solar farms, 129
solar irradiance, 96, 98, 182
solar panels, 116–117, 129, 184
solar systems, 15–16, 17
sound waves, 60–61
southern lights, 5–6
speed of light, 12–13, 62
stars, 10–11, 13–14, 14–17
steam engine, 46–47, 50, 51, 185
storms, 89–93, 155
subduction, 38, 185
sucrose, 181
sun, 10–11, 13, 59–60, 96
SunChips, 111–112
super grid, 127, 128, 130, 132, 134, 143, 162, 185
supernova, 14–15, 185
sustainable development, 156–159, 163, 186
Suzuki, David, 160
switchgrass, 139

T

tar sands. *See* oil sands
tectonic plates, 36–38, 98, 186
television, cable, 127–128

temperatures, increasing, 90–91
transient ischemic attacks (TIAs), 150
treaties, international, 144–147
turbines, 50, 118, 186
two-way communication technology, 128

U

ultraviolet radiation, 64, 186
universe, age of, 17
uranium, 15, 16–17, 136–137
utilitarianism, 153
utility companies, 121
utility quotients, 152–153

V

vector, 93, 186
vehicles, 112–113, 128–129
velocity, 13
Venus flytrap, 31
visible light, 62–65
volcanic activity, 32, 51
Volta, Alessandro, 48
Voyager 1 and *2*, 169–170

W

water
    described, 21–22, 25
    cycle of a water molecule, 22–24
    vapour, 56, 67, 68–69, 72
Watt, James, 46, 185
wave theory, 60–62
weather, severe, 89–93, 155
West Nile virus, 95
white dwarfs, 14
wind power, 117–118, 131–132
winter storms, 91, 155
wood-burning, 52–53, 70, 138

X

X-rays, 65

About the Author

BRADLEY J. DIBBLE, MD is a cardiologist living in Midhurst, Ontario. Because of his skills in bringing complex topics into an easily understandable format, he frequently gives lectures to both the public and the medical community and is a regional spokesperson for the Heart and Stroke Foundation of Ontario. His interests in the environment and in educating the public led to his appointment by the federal minister of the environment to the Sustainable Development Advisory Council, one of twenty-five members who advise on policy that affects all of Canada.

Brad is married to Katherine, who is a freelance journalist, and has two sons, Matthew and Jamie. He can be reached at b.dibble@rogers.com.

CPSIA information can be obtained at www.ICGtesting.com
Printed in the USA
LVOW090812121111
254601LV00002B/1/P